CONSTELLATIONS

CONSTELLATIONS

GILES SPARROW

CONTENTS

ORIGIN OF THE CONSTELLATIONS

Our planet is surrounded by a Universe that stretches for billions of light years in every direction. Wherever we look, stars, gas and dust clouds, and distant galaxies are scattered more or less at random across space. For thousands of years, constellations have played a key role in helping us make sense of this apparent chaos.

Today's system of 88 constellations has evolved since prehistoric times: the earliest depiction of Taurus, the Bull, comes from a 17,000-year-old painting at Lascaux, France. The first lists of constellations ancestral to our own, meanwhile, were compiled in ancient Mesopotamia more than 4,000 years ago.

Probably the earliest constellations to be recognized were those of the zodiac—a group of a dozen star patterns traditionally attributed special significance because the Sun's annual path around the sky passes through them. The word "zodiac" comes from a Greek term meaning "circle of animals," and indeed all but one of these 12 constellations represents a living creature. The exception, Libra, was once part of neighboring Scorpius. Thanks to the almost flat alignment of their orbits around the Sun, the planets are also usually found within these constellations—the idea that the positions of these "wandering" celestial bodies in the zodiac had some relationship to events on Earth dates back to at least the first millennium BC, and survives in the persistent fascination of

astrology. Thanks to long-term cycles of change in Earth's axis of rotation, however, the Sun's present-day path across the sky is largely out of step with the astrological zodiac.

While different cultures around the world developed their own star patterns at different points in history, and many of these are still in local use, astronomers today use a standardized list developed from the work of the Greek-Egyptian astronomer and geographer Ptolemy of Alexandria. Around AD 150, Ptolemy summarized the astronomical knowledge of the classical world in a great work known as the *Almagest*. Ptolemy's list of 48 constellations includes all of the ancient zodiac signs, and most of the others still recognized in northern and equatorial skies—it survived unaltered for around 1,400 years.

It was only from the 16th century that European explorers returned the first details of the far southern stars. Astronomers soon began to introduce these discoveries into their charts, often making further additions to northern skies at the same time. A dozen new constellations were added to the southern sky by the followers of Dutch navigators Pieter Dirkszoon Keyser and Frederick de Houtman around 1600. However, large areas of the sky remained forsaken until the French astronomer Nicolas Louis de Lacaille invented a further 14 constellations for his great survey of the southern stars, published in 1763.

Finally, between 1922 and 1930, the International Astronomical Union agreed a list of 88 constellations, each formally defined as an area of the sky rather than a group of stars. Together, these 88 encompass the entire celestial sphere, ensuring that every object in the sky now belongs to one constellation or another.

Within the constellations, astronomers have ordered stars in various ways. Traditionally, Greek letters are used to indicate the brightest stars, according to the system devised by German astronomer Johann Bayer in 1603. Fainter stars are often assigned "Flamsteed numbers" (introduced by English astronomer John Flamsteed in the early 18th century, while unusual stars and other celestial objects (often collectively known as "deep-sky" objects) bear numbers or letters from a variety of other systems.

The brightness of stars, meanwhile, is measured by their "apparent magnitude"—the brighter a star appears, the lower its magnitude number. This system originated in ancient Greece, but was formalized in the 19th century: today a difference of one magnitude between stars indicates that one star is roughly 2.5 times brighter than the other. The brightest stars have negative magnitudes, and the faintest naked-eye stars are around magnitude 6.0. On the maps used throughout this book, stars of different sizes are used to represent different brightnesses—more information about intepreting the maps is supplied overleaf.

This book offers a comprehensive introduction to understanding the individual constellations, and the night sky as a whole. It is divided into four sections, which combine to provide a complete guide to the celestial realm. *The Changing Heavens* describes how our unique location on Earth affects our view of the Universe, and how astronomers impose an ordered system of coordinates and constellations onto the sky. *The Fixed Constellations* takes us on a journey through the 88 officially recognized star patterns, describing the wonders that lie within them that can be seen with the naked eye, binoculars or relatively modest amateur telescopes. *The Wandering Planets* explores the Solar System interlopers that roam across the sky and alter the familiar appearance of the constellations, while *The Monthly Skies* provides a detailed guide to the skies of both hemispheres as they change their aspect on Earth's annual journey around the Sun.

The constellations were the first tools of astronomy, and still offer an important means of ordering the cosmos. Even today, an ability to recognize these celestial patterns and a knowledge of their bright stars and significant objects is a vital tool in the arsenal of any amateur stargazer. Only learn to understand the constellations, and the entire book of the Universe will open up before you.

HOW TO USE THE STAR CHARTS

On the maps used throughout the book, stars of different sizes are used to represent different brightnesses. The color of stars on the maps represents their actual measured colors, usually indicative of their surface temperature.

The brightness of stars is measured according to a system known as magnitude—the lower a star's magnitude number, the brighter it is. The key shown at upper right represents the scales of stars used in the large individual maps in *The Fixed Constellations* chapter (slightly smaller stars are used for practical purposes on maps covering wider areas in *The Changing Heavens* and *The Monthly Skies*).

On the individual maps throughout *The Fixed Constellations* section, non-stellar "deep-sky" objects are also represented using the special symbols shown in the lower key. While the stars shown on the maps should all be visibe to the naked-eye under dark skies, most of the deep-sky objects will require either binoculars or a small telescope.

KEY ►

Star magnitudes: Brighter than 0 · Brighter than 1 · Brighter than 2 · Brighter than 3 · Brighter than 4 · Brighter than 5 · Brighter than 6

Deep-sky objects: Globular cluster · Planetary nebula · Open cluster · Diffuse nebula · Spiral galaxy · Eliptical galaxy · Irregular galaxy · Other items of interest

THE GREEK ALPHABET

The brightest stars in each constellation are represented by a system of Greek "Bayer letters," in accordance with the Greek alphabet as shown here. However, the system has many inconsistencies, with some stars bearing the "wrong" letters due to a variety of historical accidents. Fainter stars down to the limits of naked-eye visibility are often given "Flamsteed numbers," linked to their location within the constellation.

α	alpha	ι	iota	ρ	rho
β	beta	κ	kappa	σ	sigma
γ	gamma	λ	lambda	τ	tau
δ	delta	μ	mu	υ	upsilon
ε	epsilon	ν	nu	φ	phi
ζ	zeta	ξ	xi	χ	chi
η	eta	ο	omicron	ψ	psi
θ	theta	ϖ	pi	ω	omega

Lighter shading indicates the extent of the Milky Way

Traditional constellation lines defining star patterns

Flamsteed numbers designate fainter stars

Other letters indicate interesting stars without a Bayer or Flamsteed designation, such as variable stars

Boundary of constellation as defined by International Astronomical Union

Greek Bayer letters designating bright stars

Grid lines show equatorial coordinate system

Deep-sky objects and their catalog numbers

Boundaries of neighboring constellations

NGC 2359

NGC 2360

θ

μ

γ

ι

SIRIUS

α

ν₃

β

MIRZAM

ν₂

π

15

M41

NGC 2207

ξ₂

ξ₁

NGC 2362

29

NGC 2354

o₂

o₁

VY

τ

δ

CANIS MAJOR

27

ω

σ

ADHARA

η

ε

ζ

κ

λ

THE CHANGING HEAVENS

FROM ANY POINT ON THE SURFACE OF THE EARTH, OUR VIEW OF THE WIDER COSMOS IS GOVERNED BY OUR PLANET'S DAILY ROTATION AND ITS ANNUAL ORBIT AROUND THE SUN. THIS CHAPTER EXPLAINS HOW SUCH CYCLES INTERACT TO CREATE OUR EVER-CHANGING NIGHT SKY AND ALSO INTRODUCES THE BROAD PATTERN OF THE CONSTELLATIONS AND THE DIFFERENT TYPES OF OBJECT WE FIND WITHIN THEM.

THE EARTH IN SPACE

OUR PLACE IN THE UNIVERSE

Ancient astronomers believed that Earth was the center of the cosmos, a fixed location in space around which all the celestial objects—the Sun, Moon, planets, and stars—moved. At first glance, this idea seems quite intuitive, but the reality of the Universe is rather different.

Many ancient peoples had models of the Universe rooted in religion, but the first cosmolgoical theory aimed at explaining the observed patterns and cycles of the heavens emerged from ancient Greece. The most successful of these involved a static Earth surrounded by the Sun, Moon, planets, and stars moving on a series of concentric spheres. Refined by the Greek-Egyptian astronomer Ptolemy around AD 150, this "geocentric" model went unchallenged for another millennium.

Then, in his 1543 work *On the Revolutions of the Heavenly Spheres*, Polish priest Nicolaus Copernicus marshalled the evidence for a "heliocentric" Universe in which Earth was just one of several planets in orbit around the Sun. Around 1610, this shift in our cosmic perspective was confirmed through the theoretical work of Johannes Kepler and the early telescopic observations of Galileo Galilei. Kepler realized that the problems with a heliocentric Universe could be resolved if the planets followed elliptical orbits rather than perfect circles.

A further diminution in Earth's stature arose from the realization that the stars were suns like our own, seen across immense distances. By the 1800s, larger telescopes allowed astronomers to begin mapping stars across our Milky Way galaxy. Today we understand that the Sun is just one insignificant star among several hundred billion, and that the Milky Way itself is one of countless similar systems across the Universe.

The immensity of cosmic distances, and the limitations of the speed with which even light can travel across space, mean astronomy is a largely observational science. Astronomers cannot travel to the stars or observe them up close, but through analysis of starlight they have still discovered a surprising amount from our obscure corner of space. Their studies have revealed the billion-year life cycles of stars, the spectacular conditions in which they are born and the strange remnants left behind when they die. They have shown the ways in which galaxies maintain their elegant structures, the processes by which solar systems and planets form, and have even offered clues to the origins and fate of the Universe itself.

But in order to get to grips with the nature of the cosmos, it is first necessary to understand the complex ways in which our location on Earth affects our view of the wider Universe.

THE VISIBLE SKY

From any point on Earth we can see roughly half of the sky—the rest is hidden by the bulk of the planet beneath our feet. The precise region of the sky we see depends on both the planet's rotation and our location on Earth and in space (see page 16).

These three diagrams show the very different skies seen by observers at Earth's north pole, at mid-latitudes and at the equator. As Earth spins on its axis, the segment of the sky visible for the mid-latitude and equatorial observers changes through the course of each day.

1. POLAR OBSERVER
An observer located at Earth's north pole sees the same hemisphere of the sky all the time, rotating daily around the north celestial pole that lies directly overhead.

2. MID-LATITUDES
From mid-latitudes, some stars rise and set as Earth's daily rotation carries different parts of the sky into view, while other "circumpolar" stars remain permanently visible.

"Circumpolar" stars visible throughout the day

"Equatorial" stars change throughout the day depending on observer's orientation

3. EQUATORIAL OBSERVER
For an observer on the equator, the celestial poles lie on opposite horizons and every region of the heavens is carried in and out of view in the course of Earth's rotation.

Direction of Earth's rotation

THE SEASONS
EARTH'S TILTED AXIS

The tilt of Earth's axis relative to the plane of the Solar System produces seasons that affect the length of day and night and the areas of the sky that we can see throughout the year. It also creates much longer-term changes in the orientation of the sky and the direction of the celestial poles.

As Earth makes its annual journey around the Sun, its axis of rotation is tilted relative to "upright" by some 23.5 degrees. The axis remains pointed in the same direction in space throughout the year and, as a result, the opposing hemispheres receive different amounts of sunlight from month to month (see opposite page).

In June, around the time of the northern "summer solstice," the northern hemisphere points toward the Sun and benefits from the Sun's longer, higher track across the sky. Meanwhile, in the southern hemisphere the days are short, the Sun's path through the sky is relatively low and its heating effect is reduced. Six months later, at northern winter solstice, the situation is reversed. Midway between these extremes, in March and September, both hemispheres receive equal amounts of sunlight at the vernal (spring) and autumnal (fall) equinoxes.

MIDNIGHT SUN

The effect of seasonal changes becomes more extreme at higher latitudes (regions around the equator, in Earth's tropical zones, receive more or less equal sunlight throughout the year). Above the Arctic and Antarctic Circles (lines of latitude at 66.5 degrees north and south), the Sun does not rise at all around the winter solstice, and never sets around the summer solstice—a phenomenon known as the midnight Sun. The north and south poles themselves experience the most extreme version of this, with six months of perpetual daylight in summer followed by an equal period of bitterly cold darkness in winter.

Orientation of Earth's
axis remains the same
throughout year

Northern summer
solstice—north pole
tilted toward the Sun

Northern winter
solstice—north pole
tilted away from the Sun

SEASONAL CHANGE

As Earth orbits around the Sun each year, the amount of
sunlight reaching different parts of Earth's surface varies
as its axis of rotation maintains the same direction in space.
The difference is at its most extreme around the solstices,
while at the intermediate equinoxes, both hemispheres
receive equal sunlight.

Equinox—both poles
equally illuminated

Changing direction of
celestial pole

Axis of
rotation

Gravitational
pull of Sun

Gravitational
pull of Moon

PRECESSION

Although Earth's tilted axis remains
pointed in the same direction from
year to year, the gravitational forces
of Sun and Moon tugging at the slight
"bulge" around our planet's equator
cause its axis to wobble in a cycle that
lasts some 25,800 years. As a result of
this cycle, known as precession, the
celestial poles describe a track around
the sky and the Sun's position at the
equinoxes tracks slowly westward.
One long-term consequence of this is
that the direction of the celestial pole
and the positions of stars in the widely
used "equatorial" coordinate system
(see page 17) change slowly over time.

THE CELESTIAL SPHERE
AN IMAGINARY SHELL AROUND THE EARTH

Although the true nature of the cosmos is very different, a simple model that imagines stars and other astronomical objects fixed on a transparent sphere around the Earth can be used as the basis for developing celestial measurement systems and mapping the skies.

The celestial sphere is an imaginary construct that flattens all the three-dimensional complexity of the Universe onto a sphere analogous to Earth itself. Instead of considering the reality of an Earth rotating on its axis once a day and spinning through space throughout the year (see page 15), this model reverts to the ancient idea of a static Earth, with a spherical shell of stars rotating around it daily, and the Sun and planets moving more slowly across this shell.

As illustrated on the opposite page, the celestial sphere pivots around points known as the celestial poles (extensions of Earth's north and south poles), and can be divided into hemispheres by a line known as the celestial equator. The Sun's annual path through the sky traces a line called the ecliptic, crossing the equator at two points corresponding to the vernal and autumnal equinoxes.

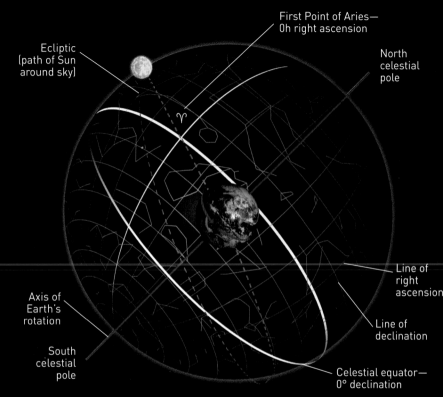

Ecliptic (path of Sun around sky)

First Point of Aries— 0h right ascension

North celestial pole

Line of right ascension

Line of declination

Celestial equator— 0° declination

Axis of Earth's rotation

South celestial pole

STARS ON A SPHERE

In this diagram, the near side of the celestial sphere has been removed for clarity. The celestial poles and celestial equator are directly aligned with Earth's own poles and equator, with latitude-like lines of declination running parallel to the celestial equator, and longitude-like lines of right ascension (R.A.) between the poles. The ecliptic (the Sun's path against the stars) is tilted at an angle to the celestial equator, crossing it in two places. The point where the Sun enters the sky's northern hemisphere at the northern spring equinox is known as the First Point of Aries (marked by the symbol ♈). It defines 0° declination, 0h right ascension.

COORDINATE SYSTEMS

Just as for locations on Earth, measuring the positions of objects in the sky requires a fixed frame of reference. The simplest way of doing this, known as the alt-azimuth system, uses an observer's local horizon. However, Earth's daily rotation means that alt-azimuth coordinates are unique to a specific location and time. In practice, "equatorial" coordinates, which measure positions relative to fixed lines on the celestial sphere, are far more useful.

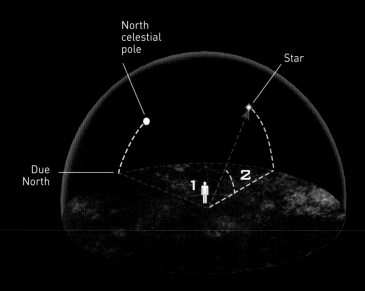

North celestial pole

Star

Due North

1

2

◄ **ALT-AZIMUTH COORDINATES**

The position of a star (or any other object) is measured according to its angle above the horizon (altitude) and its direction clockwise from due North (azimuth). An object's alt-azimuth location changes depending on the observer's location on Earth and the time of night they are observing.

1 Azimuth (angle clockwise from due North)

2 Altitude (angle above horizon)

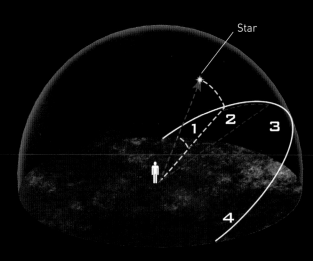

Star

1

2

3

4

◄ **EQUATORIAL COORDINATES**

In this system, a star's position is described in terms of its declination (its angle in degrees north or south of the celestial equator) and its right ascension or R.A. (measured in hours, minutes, and seconds east of 0h right ascension as defined by the First Point of Aries).

1 Declination (angle from celestial equator)

2 Right ascension (angle east of First Point of Aries)

3 First Point of Aries

4 Celestial equator

DEFINING CONSTELLATIONS
PATTERNS IN THE SKY

The constellations started off as simple patterns in the sky, made by joining up the brightest stars. But as new techniques revealed a wealth of new astronomical objects, that definition has had to change.

Constellations made simply by linking bright stars raised a few obvious issues. What if people disagreed about the shape of the pattern and the stars included? Did nearby stars outside of the pattern itself count as part of the constellation? And what about stars that were discovered later? As astronomical instruments improved, the situation was complicated by the need to catalog more and more individual objects—unique names were all very well for the brightest stars, but impractical for every star in the sky. In 1603, German astronomer Johann Bayer introduced a systematic scheme, listing stars with a combination of a Greek letter (normally corresponding to their relative brightness) and the possessive form of the constellation's Latin name. So Sirius, the brightest star in Canis Major, becomes Alpha Canis Majoris. Later catalogs attempted to number stars by brightness or position, but every scheme has its failings, and in practice celestial objects are named in a variety of different ways.

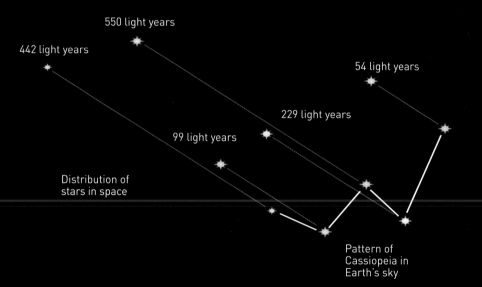

550 light years

442 light years

54 light years

229 light years

99 light years

Distribution of stars in space

Pattern of Cassiopeia in Earth's sky

A DIFFERENT POINT OF VIEW
The pattern of stars in a constellation is usually nothing more than a line-of-sight effect—the result of a chance alignment of objects, often at widely varying distances from Earth. For example, the five bright stars that form the distinctive "W" of Cassiopeia actually lie (from east to west) 442, 99, 550, 229, and 54 light years from Earth respectively.

The brightness of a star seen from Earth, called its "apparent magnitude," depends on both its true brightness or "luminosity," and its distance (dwindling according to an "inverse-square law" relationship). In other words, if two stars of the same apparent magnitude lie at 10 and 20 light years from Earth, the more distant must be four times as luminous.

A CHANGING CONSTELLATION

The stars of Taurus are one of the most recognizable patterns in the sky, identified as the form of a bull by many cultures around the world (top right). In the traditional "star pattern" interpretation of constellations (bottom left), bright stars are linked by lines and given Greek letter or number designations. In the official system of the International Astronomical Union, however (bottom right), the constellation is a region of the celesial sphere and includes everything contained within it.

OBJECTS IN THE SKY

PLANETS, STARS, NEBULAE, AND GALAXIES

Telescopic views of the sky reveal a variety of different objects, ranging from solid and gaseous planets to fiery stars, tenuous nebulae and great, slowly rotating clouds and spirals of stars.

At a casual glance, lights in the night sky might appear to be made of nothing but stars—but look in more detail, even with the naked eye, and differences soon emerge. The planets appear as tiny disks rather than star-like pinpricks of light, and are not luminous bodies in their own right, shining instead through reflected sunlight. Planets are relatively small objects in the cosmic scheme of things, born out of the leftover material in orbit around recently formed stars.

Stars themselves are huge and luminous—incandescent balls of gas that are powered by nuclear fusion, transforming light elements into heavier ones in their cores. A star's color reveals its surface temperature—cooler stars appear red, while hotter ones are blue. Our yellow-white Sun is a fairly average star with a surface temperature of 9,900 °F (5,500 °C). Stars also vary hugely in brightness—the smallest "red dwarfs" may be 10,000 times fainter than the Sun, while the brightest giants (typically blue or red in color) can be 100,000 times brighter. Depending on a star's mass, it may live for billions or merely millions of years, and end its life shrugging off its outer layers as a "planetary nebula" or tearing itself apart in a spectacular supernova explosion. Star death leaves behind a stellar remnant: either a low-mass "white dwarf" star, a heavier neutron star or an extremely massive black hole.

Stars are born in nebulae—huge clouds of gas and dust that lie between the stars. Normally, these are invisible except when they are silhouetted against a brighter background and known as dark nebulae. However, when something—perhaps a supernova shock wave or the tides raised by a passing star or the rotation of a galaxy—triggers a process of collapse, the clouds can separate into rapidly condensing protostars, which begin to shine when they become hot and dense enough to trigger fusion. As young stars emerge from their cocoons in tight groups known as clusters, their fierce radiation can reflect from nearby gas or cause it to fluoresce, producing beautiful reflection and emission nebulae.

On the largest scales, galaxies are huge agglomerations of stars, grouped together in clusters and superclusters. Complex and beautiful spirals, and chaotic but bright irregular galaxies, are rich in star-forming material and are therefore dominated by brilliant, massive but short-lived young stars. Ball-shaped elliptical galaxies, in contrast, lack star-forming material and are therefore dominated by longer-lived red and yellow stars.

COSMIC SPECTACLE

Jupiter (top left) is the largest planet in the Solar System—even the smallest telescope will reveal the light and dark bands of its enormous weather systems. Open clusters, such as NGC 3603 in Carina, are dominated by brilliant heavyweight stars (top right), and are sometimes still surrounded by the nebulosity from which they formed. The Trifid Nebula in Sagittarius, meanwhile, combines dark dust lanes, pinkish emission regions, and bluish areas of reflection (bottom left). Spiral galaxies, such as the interacting pair Arp 273 (bottom right), are mostly too distant to identify their individual stars, but are nevertheless defined by the light from their brightest nebulae and star clusters.

MAPPING THE HEAVENS

CIRCUMPOLAR SKIES

The series of maps on the
following pages depicts
the constellations in the
wider context of the
celestial sphere. For
practical purposes,
this sphere can
be divided into
six segments—
circumpolar stars
around each celestial
pole, and four separate
quadrants stars to
either side of the celestial
equator, centered on
specific lines of right
ascension. The maps on
these pages shows
the northern and southern
circumpolar stars

Northern
circumpolar
constellations
include well-kn
patterns such a
Ursa Major and
Cassiopeia, but
also more obsc
regions, such
as Cepheus and
Camelopardalis
north celestial
itself is marked
the Pole Star P
in Ursa Minor (a
center of the m

13H
12H
11H
14H
10H
CENTAURUS
VELA
9H
15H
LUPUS
8H
16H
CRUX
TRIANGULUM
AUSTRALE
NORMA
MUSCA
7H
APUS
CHAMAELEON
CARINA
17H
CIRCINUS
VOLANS
ARA
ÖCTANS
-80°
6H
18H
-70°
-60°
-50°
-40°
PAVO
MENSA
PICTOR
TELESCOPIUM
DORADO
19H
HYDRUS
RETICULUM
5H
TUCANA
HOROLOGIUM
20H
4H
INDUS
PHOENIX
3H
21H
GRUS
22H
2H
23H
0H
1H

SOUTHERN HEMISPHERE

The southern circumpolar skies are filled with fairly obscure constellations such as Hydrus, Pavo, and Volans, while the south celestial pole lies in Octans, with no nearby bright star to mark it. A circle of brighter stars surrounds the pole at a distance of about 30 degrees.

This pair of charts shows the equatorial regions of the sky centered around right ascension coordinates of 18h and 12h (see page 16). The stars in these regions of the sky are visible from the middle latitudes of both hemispheres at different times of year and different times of night. Stars around 18h rise around 6 p.m. local standard time (LST) during June, while those around 12h rise around 6 p.m. LST in March.

The stars around 18h right ascension are familiar sights on northern summer (southern winter) nights. They run along a broad and bright swathe of the Milky Way, from Cygnus and Lyra to the north, through Ophiuchus and Serpens, to Sagittarius and Scorpius in the south.

12H RIGHT ASCENSION

The stars around 12h right ascension dominate the evening skies during northern spring and southern fall. They include the large and umistakable pattern of Leo, the bright stars Arcturus in Boötes and Spica in Virgo, the enormous but faint Hydra, and the upper reaches of Centaurus.

These two maps show the equatorial skies centered around 6h and 0h right ascension. Stars in these areas of the sky are visible from the mid-latitudes of both hemispheres at different times throughout the year. Stars around 6h rise around 6 p.m. local standard time (LST) during December, while those around 0h rise around 6 p.m. LST in September.

The constellations around 6h right ascension are prominent on northern winter (southern summer) evenings. They include the bright stars of Orion and its retinue, including Canis Major and Canis Minor around the celestial equator, and Taurus, Gemini, and Auriga to the north. South of Orion lie the long river Eridanus and the bright stars of Puppis and Vela.

0H RIGHT ASCENSION

The stars around 0h right ascension are best seen in evening skies during northern spring and southern fall. The bright Square of Pegasus dominates the northern reaches of the sky, with Andromeda, Triangulum and Pisces nearby, while Cetus and Capricornus lie close to the equator. Piscis Austrinus, with its bright star Fomalhaut, is the most prominent of the fainter southern constellations.

THE FIXED CONSTELLATIONS

EIGHTY-EIGHT CONSTELLATIONS DIVIDE THE ENTIRE
CELESTIAL SPHERE INTO EASILY MANAGEABLE REGIONS.
THE OLDEST HAVE ORIGINS THAT ARE LOST IN THE
MISTS OF PREHISTORY, WHILE THE YOUNGEST ARE
INVENTIONS OF THE 18TH-CENTURY ENLIGHTENMENT.
THIS CHAPTER EXPLORES THE CONSTELLATIONS AND
THE OBJECTS THAT ARE FOUND WITHIN THEM IN DETAIL.
THROUGHOUT THE CHAPTER, CONSTELLATIONS ARE
ARRANGED IN APPROXIMATE AVERAGE DECLINATION
FROM THE NORTHERNMOST TO THE SOUTHERNMOST.

URSA MINOR
THE LESSER BEAR

The northernmost constellation in the sky, Ursa Minor is famed for its brightest star, Polaris, the Pole Star. Polaris lies within half a degree of the north celestial pole and remains almost perfectly still in the sky even as the rest of the heavens spin around it.

Ursa Minor consists of seven moderately bright stars whose pattern strongly resembles the larger and brighter shape of the Big Dipper, or Plough, in the nearby Great Bear Ursa Major (see page 44). This resemblance is probably the reason why the constellation eventually became known as the Lesser Bear, though others have seen it as a hunting dog or even an extended wing of the mighty dragon Draco. Polaris, the constellation's brightest star with a magnitude of 2.0 and a distance of some 2,400 light years, is a rare example of a star that has changed noticeably in recorded history. It is an entire magnitude brighter than it was in ancient times and until recently was a pulsating variable star— its changes in size and brightness have only dwindled away over the past century.

POLAR TRAILS Long-exposure photographs of the region around the north celestial pole capture the sky's apparent rotation over a period of up to an hour. All the stars in the sky describe arcs of light as Earth spins beneath them, but the arcs are shortest for those stars closest to the pole, including Polaris.

	R.A.	DEC.	TYPE	MAGNITUDE	DISTANCE
Polaris (α)	02h 31m	+89°15''	Binary star	2.0	2,400 l.y.
Pherkad (γ)	15h 20m	+71°50''	Variable star	c.3.1 (var)	540 l.y.
η UMi	16h 18m	+75°45''	Visual double	5.0, 5.5	97,650 l.y.

DRACO
THE DRAGON

The long, sinuous figure of Draco, the celestial dragon, winds its way around the north celestial pole, all but enclosing the smaller Ursa Minor. Yet despite its size, Draco is disappointingly lacking in deep-sky objects and bright stars.

In mythology, Draco was the dragon that guarded the golden apples in the orchard of the Hesperides, and was fought by the great hero Hercules as one of his 12 tasks. In today's skies, Hercules still kneels above the dragon's head, club raised to deliver a killing blow.

The constellation's brightest star is Eltanin, or Gamma Draconis, an orange giant some 150 light years from Earth that shines at magnitude 2.2. It happens to be heading toward Earth at a speed of around 17.5 miles per second (28 kilometers per second), and will pass within 28 light years of Earth around 1.5 million years from now, becoming the brightest star in the sky. Magnitude-3.6 Alpha Draconis, or Thuban, meanwhile, is a tight binary that, thanks to the effects of precession (see page 15) was pole star around 2,800 BC.

CAT'S EYE NEBULA Draco's most celebrated object is the planetary nebula NGC 6543, also known as the Cat's Eye Nebula. Lying some 3,600 light years from Earth and faintly visible through a small telescope, the Cat's Eye is formed from glowing shells of gas cast by a giant star passing through a period of instability at the end of its life. This Hubble Space Telescope image reveals intricate spiral structures created as stellar winds blowing out from the central star catch up with and "inflate" material cast off centuries before.

	R.A.	DEC.	TYPE	MAGNITUDE	DISTANCE
Thuban (α)	14h 04m	+64°22"	Binary star	3.6	310 l.y.
Arrakis (μ)	17h 05m	+54°28"	Binary star	4.9	90 l.y.
39 Dra	18h 24m	+58°48"	Multiple star	5.0	190 l.y.
NGC 6543	17h 58m	+66°38"	Planetary nebula	8.1	3,600 l.y.

CEPHEUS
THE KING OF ETHIOPIA

The brightest stars of this high northern
constellation form a pattern resembling a
child's drawing of a house, and are easily
ignored among their more obvious and
prominent neighbors. Yet with a position on
the borders of the northern Milky Way,
Cepheus has several overlooked treasures.

Cepheus represents the King of Ethiopia—husband
of Cassiopeia and father of princess Andromeda in
the Perseus legend (see page 74). The constellation's
main claims to fame are its variable stars—the
prototypes for no less than three different forms of
pulsating stars are found here. Beta Cephei (Alfirk)
is the brightest of these, a blue-white giant star
some 595 light years from Earth whose variations of
roughly 0.1 magnitude every 4.6 hours are too small
and rapid a fluctuation to be seen without specialist
equipment. Delta Cephei, in contrast, varies between
magnitudes 3.5 and 4.4 every 5.37 days. Its changes
are easily spotted by comparison with neighboring
stars, and it is the prototype for the class commonly
known as "Cepheids." which astronomers use to
measure the distance of other galaxies.

GARNET STAR NEBULA A large nebula cataloged as IC 1396 lies close to
the Garnet Star Mu Cephei, one of the most celebrated stars in the sky. Shining at
magnitude 4.0 across 5,000 light years, Mu is a red supergiant, emitting as much
energy as 350,000 Suns and varying unpredictably in brightness. IC 1396, some 2,000
light years closer to Earth, is rich in detail such as the dark dust lanes shown here.

	R.A.	DEC.	TYPE	MAGNITUDE	DISTANCE
Alfirk (β)	21h 29m	+70°34"	Variable star	3.15–3.2	595 l.y.
δ Cep	22h 29m	+58°25"	Variable star	3.5–4.4	890 l.y.
Garnet Star (μ)	21h 44m	+58°47"	Variable star	3.6–5.0	5,000 l.y.
ο Cep	23h 18m	+68°07"	Binary star	4.9/7.1	211 l.y.

CAMELOPARDALIS
THE GIRAFFE

This indistinct constellation, covering a large swathe of otherwise vacant sky, is a comparatively late addition to the northern heavens, and supposedly represents a giraffe, although it is hard to identify any real pattern among its faint stars.

Camelopardalis was introduced to celestial charts by the Dutch astronomer and theologian Petrus Plancius around 1600. Plancius apparently intended it to represent the camel that carried Rebecca to her wedding with Isaac in the Bible's Old Testament, but he mistakenly used the Latin word for a giraffe, and the name stuck.

The constellation's main objects of interest are two unpredictable variable stars, Beta and Z Camelopardalis. Beta is a yellow supergiant of magnitude 4.0 that is prone to unexpected surface flares: in 1967 it jumped in brightness by a whole magnitude for just a few minutes. Z Cam is a "dwarf nova"—a double star system, normally of magnitude 13, which increases its brightness by roughly two magnitudes every few weeks, thanks to enormous explosions in the atmosphere of one of the stars.

GALAXY NGC 2403 The brightest galaxy in Camelopardalis lies 12 million light years away and shines at magnitude 8.4. However, because it is almost face-on to Earth its light is spread out, and it is best seen in dark skies using binoculars or a low-powered telescope. NGC 2403 forms part of the same small galaxy group as the better known Messier 81 and 82, over the border in Ursa Major (see page 44).

	R.A.	DEC.	TYPE	MAGNITUDE	DISTANCE
β Cam	05h 03m	+60°27''	Multiple star	4	1,000 l.y.
Z Cam	08h 25m	+73°07''	Dwarf nova	10.0–14.5	530 l.y.
NGC 1502	04h 08m	+62°20''	Open cluster	6.9	3,100 l.y.
NGC 2403	07h 37m	+65°36''	Spiral galaxy	8.4	12 million l.y.

CASSIOPEIA
THE QUEEN OF ETHIOPIA

This distinctive W-shaped constellation is a fixture of far northern skies, circling around the north celestial pole opposite the equally distinctive Big Dipper or Plough. Cassiopeia represents the queen of the same name in the Perseus legend, the famously vain wife of King Cepheus and mother of the princess Andromeda.

The constellation wheels high overhead on northern summer nights, and is home to the rich starfields of the northernmost Milky Way. Deep-sky objects include the open clusters M52, M103 and NGC 457, and the supernova remnant Cassiopeia A.

Alpha Cassiopeiae, or Schedar, is a yellow-orange giant star, 230 light years from Earth and 500 times more luminous than the Sun. Nineteenth-century astronomers considered it to be a variable star, but today it shines at a steady magnitude 2.25. Iota Cassiopeiae, meanwhile, is a multiple star system—with a magnitude-4.5 primary star, a magnitude-8.4 companion visible through small telescopes and a further component of magnitude 6.9, only resolvable through larger instruments.

CASSIOPEIA A A tattered, expanding bubble of superhot gas, barely visible at optical wavelengths but a strong source of both X-rays and radio waves, marks the site of the Milky Way galaxy's most recent known supernova explosion. This particular supernova, marking the death of a massive star 11,000 light years away, should have been seen on Earth around 300 years ago, but apparently went unnoticed at the time.

	R.A.	DEC.	TYPE	MAGNITUDE	DISTANCE
γ Cas	00h 47m	+60°43"	Variable star	2.2–3.4	610 l.y.
ι Cas	02h 29m	+67°24"	Multiple star	4.5/8.4/6.9	141 l.y.
M52	23h 24m	+61°35"	Open cluster	7.3	5,000 l.y
M103	01h 33m	+60°42"	Open cluster	7.4	8,500 l.y
NGC 457	01h 19m	+58°20"	Open cluster	6.4	9,000 l.y

AURIGA AND LYNX

THE CHARIOTEER AND THE LYNX

Auriga is a prominent constellation in mid-northern skies, dominated by the brilliant star Capella, and home to several interesting stars and clusters. Its near neighbor Lynx, in contrast, is renowned for its faintness and obscurity.

Although Auriga has been seen as a charioteer since ancient times, its precise identity varies—the figure may be one of two legendary Greek heroes, Eritchthonius or Myrtilus. The constellation's most prominent stars, meanwhile, are associated with a completely different story—Capella's name means "she goat," and the star represents Amaltheia, the goat that suckled the infant Zeus. The tight triangle of stars just to the southwest of Capella, meanwhile, are known as "the kids." Capella itself is a complex quadruple star system 42 light years from Earth, though it takes a moderate telescope to spot even the brightest of the primary star's companions.

In stark contrast, Lynx is a 17th-century invention of Polish astronomer Johannes Hevelius, who filled a gap in the sky with a chain of stars so faint, he said, that only the lynx-eyed could spot it

AE AURIGAE At the limits of naked-eye visibility, AE Aurigae is a seemingly unremarkable blue star around 1,400 light years from Earth. Its radiation illuminates the faint but beautiful structure of IC 405, the so-called Flaming Star Nebula, but the star did not originate in this region of space—instead it is a rare, fast-moving runaway, ejected from its birthplace in the Orion Nebula around 2 million years ago.

	R.A.	DEC.	TYPE	MAGNITUDE	DISTANCE
Capella (α)	05h 17m	+46°0"	Multiple star	0.1	42 l.y.
ε Aur	05h 02m	+43°49"	Variable star	2.9–3.8	2,000 l.y.
ζ Aur	05h 02m	+41°05"	Binary star	3.8	790 l.y.
12 Lyn	06h 46m	+59° 27"	Multiple star	5.3/6.2/7.2	230 l.y.
M38	05h 28m	+35°50"	Open cluster	7.4	4,200 l.y.
NGC 2419	07h 38m	+38°53"	Globular cluster	10.4	295,000 l.y

EPSILON (ε) AURIGAE

The northernmost of "the kids" triangle of stars, Epsilon Aurigae is a remarkable variable star whose true nature has only recently been established. It normally shines at a steady magnitude 3.0, but every 27 years its brightness drops gradually to magnitude 3.8, where it remains for about a year before recovering. This behavior naturally suggests an eclipsing binary star, but the eclipsing secondary object seems to be semitransparent and makes no contribution to the system's light at other times. What's more, the length of the eclipses suggests it must be around 3 billion km (1.9 billion miles) across. One long-standing theory to explain this strange behavior is that the eclipse is caused by a disk of planet-forming material in orbit around an unseen object that is itself orbiting the primary star. This was finally confirmed during the 2009–11 eclipse, when astronomers imaged the disk crossing the face of Epsilon Aurigae for the first time.

R.A. 05h 02m, Dec. +43°49"
Magnitude 2.9–3.8 (var)
Distance 2,000 light years

THE INTERGALACTIC WANDERER NGC 2419

This faint globular cluster, only visible through medium-sized telescopes, stands out from others of its type for two main reasons. Firstly, it lies on the opposite side of the sky from most other globular clusters that orbit around the core of the Milky Way—there are no other globulars within 60 degrees of it. Secondly, and even more remarkably, it is almost 300,000 light years from Earth—much further away than our galaxy's other globular clusters, and even more distant than some of the Milky Way's satellite galaxies. It is currently approaching us at a speed of 12.5 miles per second (20 km per second), but astronomers suspect it is not gravitationally bound to the Milky Way at all—it may be an outcast from another galaxy altogether, flung off into intergalactic space during a cosmic close encounter.

R.A. 07h 38m, Dec. +38°53"
Magnitude 10.4
Distance 295,000 light years

▶ IC 405

FLAMING STAR NEBULA

This beautiful nebula around the speeding runaway star AE Aurigae is illuminated by a mixture of emission and reflection processes. Visible light from the central star is reflected and scattered by fine dust particles within the nebula, which tend to bend shorter wavelengths more than longer ones and therefore give the light deflected toward Earth a blue tint. Meanwhile, invisible ultraviolet radiation energizes the nebula's gas atoms and molecules, causing them to emit radiation at visible wavelengths as they return to their normal state.

R.A. 05h 16m, Dec. +34°27″
Magnitude c.6.0 (var)
Distance 1,400 light years

URSA MAJOR
THE GREAT BEAR

While the Big Dipper or Plough is the most instantly recognizable of star patterns for northern hemisphere observers, the constellation of the Great Bear extends across a much larger region.

Ursa Major is the third-largest constellation in the entire sky, and has been imagined as a bear by widely separated civilizations since ancient times. Its seven brightest stars form the body and tail of the bear, while fainter outlying stars mark the head and limbs. The Big Dipper is a famously useful celestial signpost—its two brightest stars Dubhe and Merak (on the front of the dipper's "pan") point directly toward the Pole Star Polaris, while the arc of its tail can be extended past Arcturus in Boötes, as far as Spica in Virgo. Many of the constellation's stars are physically close to each other in space, around 80 light years from Earth, and moving in the same direction through space—indications that they all originated in the same star cluster. Among the Plough stars, only Alpha Ursae Majoris (Dubhe) and Eta (Alkaid) are not members of the so-called "Ursa Major Moving Group."

PINWHEEL GALAXY The large face-on spiral galaxy Messier 101 forms a rough triangle with Mizar and Alkaid, the two easternmost stars of the Big Dipper's handle. Although ostensibly quite bright at magnitude 7.9, its light is diffuse because it happens to lie face-on as seen from Earth. As a result, it is best observed through binoculars or a telescope at low power, where it appears as a pale patch of light about half the size of the full Moon. The Pinwheel lies some 27 million light years away.

	R.A.	DEC.	TYPE	MAGNITUDE	DISTANCE
Dubhe (α)	11h 04m	+61°45"	Binary star	1.8	124 l.y.
Mizar (ζ)	13h 25m	+54°56"	Multiple star	2.3	78 l.y.
M81	09h 56m	+69°04"	Spiral galaxy	6.9	12 million l.y.
M82	09h 56m	+69°41"	Irregular galaxy	8.4	12 million l.y.
M1	h 03m	+54°21"	Spiral galaxy	7.9	27 million l.y.

MIZAR AND ALCOR
ζ URSAE MAJORIS

The middle star of the Bear's tail, Zeta Ursae Majoris or Mizar, is a renowned multiple star. Binoculars or even sharp eyesight reveal a magnitude-4.0 companion, Alcor, close to magnitude-2.3 Mizar, and even a small telescope will show that Mizar is itself binary. What's more, each of these three stars is a double star in turn. For a long time, the grouping of Mizar and Alcor was seen as mere chance, but new studies have suggested that the stars are genuinely bound by gravity, creating a sextuple system.

HUBBLE DEEP FIELD

Over ten days in December 1995, astronomers used the Hubble Space Telescope to take a series of images of a small patch of apparently empty sky in Ursa Major. Combining the multiple exposures using computerized image processing, NASA scientists produced the Hubble Deep Field (HDF)—a view that reveals some 3,000 galaxies stretching across billions of light years of space and offers a new perspective on the Universe.

R.A. 13h 25m, Dec. +54°56''
Magnitude 2.3, 4.0
Distance 78 light years

R.A. 12h 37m, Dec. +62°12''
Magnitude < 28
Distance Up to 12 billion light years

◀ BODE'S GALAXY
MESSIER 81

This tightly wound spiral galaxy,
12 million light years from Earth,
bears the name of Johann Elert
Bode, the German astronomer who
discovered it in 1774. Binoculars
show it as a fuzzy point of light
while a small telescope will reveal
the oval shape of its central nucleus
and larger instruments may trace
its spiral arms. Although most of
the light from its central regions is
due to sheer density of stars, M81
also emits some radiation directly
from its nucleus, making it one
of the closest "active galaxies". It
forms the core of one of the closest
galaxy groups to our own.

R.A. 09h 56m, Dec. +69°04"
Magnitude 6.9
Distance 12 million light years

47

URSA MAJOR THE CIGAR GALAXY M82

MULTIWAVELENGTH COMPOSITE

This colorful image combines data from visible-light, infrared and X-ray telescopes to produce a unique view of the Cigar Galaxy. Infrared observations from the Spitzer Space Telescope are shown in red, a visible-light image from the Hubble Space Telescope in white, and X-ray data from the Chandra X-ray Observatory in blue. The composite shows how X-ray emissions are concentrated in plumes that emerge from above and below the galaxy's nucleus—they are mostly "synchrotron radiation" generated by charged electron particles moving at high speeds after their expulsion from the galaxy's active core.

STARBURST CORE

A detailed Hubble view of the Cigar's central regions reveals brilliant, compact star clusters embedded in a dusty environment. Each of these "super star clusters" contains 100,000 stars or more—they are thought to be the infant stage of the globular clusters found in orbit around our own galaxy. Their formation is triggered by violent galactic encounters, in this case a glancing collision between M82 and the nearby spiral Messier 81 (see page 47). The properties of stars within the clusters suggest that they were created in a 100-million-year "starburst" that began around 600 million years ago. The repercussions of this event are still being felt as disrupted material continues to fall onto the galaxy's central supermassive black hole, fueling its ongoing activity.

▶ AN EXPLODING GALAXY

A popular target for amateur stargazers, the Cigar Galaxy is one of our brightest galactic neighbors, with a compact structure that shines at magnitude 8.4. Its elongated shape, curious appearance and bright central regions led to its classification as an irregular galaxy, but in 2005 astronomers discovered traces of a spiral structure. This Hubble Space Telescope image combines light taken with four filters including visible and infrared light, and shows the emissions from filaments of hydrogen that surround the galaxy and give rise to its explosive appearance.

R.A. 09h 56m, Dec. +69°41"
Magnitude 8.4
Distance 12 million light years

CANES VENATICI

THE HUNTING DOGS

Although it has only one bright star to its name, the constellation of the Hunting Dogs is easily located thanks to its position between the familiar Big Dipper and the bright star Arcturus in Boötes. Its apparent emptiness belies the presence of several interesting objects.

Canes Venatici was added to the sky by Polish astronomer Johannes Hevelius in the late 1600s—prior to this time Arab astronomers had associated the area with the crook of the shepherd Boötes. The brightest star, Cor Caroli or "Charles's Heart," was named by British astronomer Edmond Halley to commemorate the executed King Charles I. At magnitude 2.9, it is an attractive double, easily split with a small telescope to reveal a companion of magnitude 5.6. A little way to the north of Beta lies Y CVn, or "La Superba," a red giant star on the verge of transforming into a planetary nebula. It fluctuates between magnitudes 4.8 and 6.3 in around 158 days. Another highlight of this small constellation is the globular cluster M3. At 34,000 light years from Earth, his rich star cluster shines at magnitude 6.2.

MESSIER 106 Tucked into the northwestern corner of Canes Venatici, M106 is an unusual spiral galaxy around 25 million light years from Earth. It was discovered in 1781 and is today classified as a Seyfert II galaxy—a type of active galaxy that has an unusually bright central region emitting radiation at a variety of wavelengths. Its strangest features are "anomalous" arms that are invisible in normal light (shown blue and purple in this composite image).

	R.A.	DEC.	TYPE	MAGNITUDE	DISTANCE
Cor Caroli (α)	12h 56m	+38°19″	Multiple star	2.9/5.6	110 l.y.
La Superba (Y)	12h 45m	+45°26″	Variable star	4.8–6.3	710 l.y.
M3	13h 42m	+28°23″	Globular cluster	6.2	34,000 l.y.
M51	13h 30m	+47°12″	Spiral galaxy	8.4	23 million l.y.
M94	12h 51m	+41°07″	Spiral galaxy	8.2	15 million l.y.
M106	12h 19m	+47°18″	Spiral galaxy	8.4	25 million l.y.

INFRARED SKELETON This pair of Hubble Space Telescope images contrast the Whirlpool in visible light (left) and near-infrared radiation (right). Most of the galaxy's stars emit their light at visible wavelengths, so the infrared image effectively filters them out, revealing instead the distribution of warm gas and dust too cool to emit visible light. A comparison reveals that, while the disk of the galaxy is filled with this material the opaque dust lanes that follow the Whirlpool's spiral structure are distinctly warmer than their surroundings. This is because they are the key locations where new stars are beginning to form, but have not yet become hot enough to shine properly. Interestingly, the galaxy's bright hub shows no clear difference from its surroundings—ongoing star formation in spirals such as the Whirlpool and our own galaxy is concentrated in the spiral arms rather than the central regions.

BRIGHT NUCLEUS

This early image from the Hubble Space Telescope zooms in on the star-like point of light at the center of the galaxy. Image processing reveals the presence of a dark cross in front of a bright light source, and for some time this was thought to indicate two dust rings encircling a central supermassive black hole. Further observations have now shown that the cross is due to light-absorbing dust in the foreground. However, the brightness of the nucleus certainly indicates the presence of a black hole—as does the pair of X-ray-emitting lobes of multi-million-degree gas, which extend to several hundred light years on either side of the nucleus.

DANCING PARTNERS

At magnitude 8.4, Messier 51 is one of the brightest galaxies in the sky despite an impressive distance of 23 million light years. Binoculars can locate its diffuse form, but a medium-sized telescope is needed to pick out its spiral arms. One of M51's spiral arms appears to link it to the nearby irregular galaxy NGC 5195, but this is a line-of-sight effect (in fact NGC 5195 lies behind the spiral arm). However, the galaxies are still close enough to affect each other—M51's gravity may have triggered a burst of star formation in its smaller neighbor, while NGC 5195's influence may have awoken M51's active nucleus and helped to reinforce its spiral pattern.

R.A. 13h 30m, Dec. +47°12"
Magnitude 8.4
Distance 23 million light years

BOÖTES
THE HERDSMAN

This constellation's distinctive kite-like shape and brightest star Arcturus make it a familiar feature of northern skies as it chases the Greater and Lesser Bears around the celestial pole.

Boötes is traditionally said to represent a celestial herdsman, driving the bears away from his flock with the assistance of his hunting dogs (nearby Canes Venatici). In another interpretation, however, he represents Arcas, son of Zeus and the beautiful Callisto, who was transformed into the Great Bear by the jealous goddess Artemis.

With a name meaning "bear follower," the constellation's brightest star, Arcturus, is the closest red giant star to Earth, just 37 light years away. It is easily located by following the arc of the three stars in the handle of the Big Dipper, and lies at the base of an elongated kite shape that forms the body of the constellation. Arcturus is thought to be a star quite similar to the Sun, but more advanced along its life cycle. Having exhausted the hydrogen fuel supply at its core, it is now advancing toward the final stages of its evolution.

TAU BOOTIS This Sunlike star, some 51 light years from Earth, is home to one of the first extrasolar planets to be discovered. Tau Boötis b has roughly four times the mass of Jupiter, but orbits its star in just 3 days 7.5 hours. Much further out, a red dwarf binary companion circles the primary star in a thousand-year orbit. Discovered in 1996, Tau Boötis b was one of the first extrasolar planets in a class known as the "hot Jupiters."

	R.A.	DEC.	TYPE	MAGNITUDE	DISTANCE
δ **Boo**	15h 16m	+33°19"	Binary star	3.5/7.8	122 l.y.
Izar (ε)	14h 45m	+27°04"	Multiple star	2.4	210 l.y.
Alkalurops (μ)	15h 25m	+37°23"	Multiple star	4.3	120 l.y.
τ **Boo**	13h 47m	+17°27"	Planetary system	4.5	51 l.y.
κ **Boo**	14h 13m	+51°47"	Multiple star	4.5/6.4/4.6	155 l.y.
ξ **Boo**	14h 51m	+19°06"	Binary star	4.7/7.0	22 l.y.

CORONA BOREALIS

THE NORTHERN CROWN

This distinctive circlet of stars, lying close to Boötes in a fairly empty region of the northern sky, makes up for a lack of deep-sky objects with interesting variable stars.

The Northern Crown is one of the 48 constellations listed by the Greek-Egyptian astronomer Ptolemy in the second century. It represents a coronet worn by the Cretan princess Ariadne at her wedding to Dionysus, the god of wine.

R Corona Borealis, near the center of the coronet, is a yellow supergiant 6,000 light years from Earth. Normally it shines at magnitude 5.9, on the limit of naked-eye visibility, but it sometimes drops to magnitude 14, beyond the reach of most amateur telescopes. Astronomers think these unpredictable plunges are triggered when the star throws large amounts of carbon into its atmosphere, where it forms dark clouds. Meanwhile, 1,800 light years away and close to Epsilon, the "Blaze Star" T Corona Borealis shows the reverse behaviour. It is a "recurrent nova" system, normally languishing at around magnitude 11.0 but prone to huge explosions that can raise it to magnitude 2.0 every few decades.

ABELL 2065 This dense galaxy cluster lies in the southwest corner of Corona Borealis, and is one of the most distant visible with amateur equipment. Despite this, its combined magnitude of around 14.0 still requires a fairly large amateur telescope or a long photographic exposure. Abell 2065 is one of many distant clusters cataloged by U.S. astronomer George Abell in the 1950s. It contains more than 400 galaxies and lies around a billion light years from Earth.

	R.A.	DEC.	TYPE	MAGNITUDE	DISTANCE
ζ CrB	15h 39m	+36°38"	Binary star	5.1/6.0	470 l.y.
σ CrB	16h 15m	+33°52"	Multiple star	5.8/6.7/10.8	71 l.y.
R CrB	15h 49m	+28°09"	Variable star	5.9–14.0	6,000 l.y.
T CrB	16h 00m	+25°55"	Dwarf nova	2.0–11.0	1,800 l.y.

HERCULES

HERCULES

This large northern constellation represents the great hero of Greek and Roman myth. Despite its size, Hercules is lacking in bright—its most prominent deep-sky objects are globular star clusters, including one of the sky's finest.

Best located by looking midway between the bright stars Vega in Lyra and Arcturus in Boötes, Hercules consists of a lopsided quadrangle of stars known as the Keystone, with extended chains of stars at each corner marking the hero's limbs. He is usually depicted upside down, kneeling with his foot on the head of the dragon Draco, and a club in one hand. In Greek myth, he was famously forced to accomplish 12 tasks to atone for killing his family.

The hero's head is represented by the double star Rasalgethi, which is easily resolved through a small telescope to reveal a bright red star of magnitude 4.5 and a greenish-white companion of magnitude 5.4. Delta Herculis is another attractive double for small telescopes, consisting of a blue primary of a magnitude 3.1, with a companion of magnitude 8.2.

MESSIER 13 Hercules is home to the finest globular cluster in the northern sky—Messier 13. This huge cluster contains roughly a million stars crammed into a ball just 150 light years across. At a distance of 25,000 light years from Earth, it is just visible to the naked eye, and appears through binoculars as a circular patch of light. Small telescopes can pick out individual chains of stars around its outer edges.

	R.A.	DEC.	TYPE	MAGNITUDE	DISTANCE
ζ Her	16h 41m	+31°36"	Multiple star	2.9/5.5	35 l.y.
M13	16h 42m	+36°28"	Globular cluster	5.8	25,000 l.y.
M92	17h 17m	+43°08"	Globular cluster	6.4	27 million l.y.
NGC 6210	16h 45m	+23°49"	Planetary nebula	9.0	6,000 l.y.

LYRA
THE LYRE

This compact constellation, instantly recognizable thanks to the presence of Vega, the fifth-brightest star in the sky, represents the ancient musical instrument played by the Greek hero Orpheus, among others. Lying close to the dense star clouds of the northern Milky Way, it is home to several interesting objects, including the famous Ring Nebula.

Vega, or Alpha Lyrae, is not simply interesting on account of its brightness. It is comparatively close—just 25 light years from Earth—and quite young at about 500 million years old (one-tenth of the age of the Sun). Infrared studies have revealed that it is still surrounded by a disk of gas and dust that may be in the process of forming planets.

Another highlight of the constellation is Epsilon Lyrae, a famous multiple star system. Binoculars show that it consists of two distinct components of magnitudes 4.7 and 4.6, but small telescopes will reveal that each of these stars is a double in its own right. Beta Lyrae or Sheliak, meanwhile, is a much tighter system—an "eclipsing binary" star.

RING NEBULA Midway between Beta and Gamma Lyrae lies Messier 57, a faint cosmic smoke ring best seen through a small telescope. The Ring is a famous planetary nebula—a complex expanding shell of gas shrugged off by a dying Sunlike star, and energized by radiation from the hot star at its center. It lies around 2,300 light years from Earth and is about 1.5 light years across.

	R.A.	DEC.	TYPE	MAGNITUDE	DISTANCE
Sheliak (β)	18h 50m	+33°22"	Variable star	3.4–4.3	950 l.y.
ε Lyr	18h 44m	+39°40"	Multiple star	4.7/6.2/5.1/5.5	160 l.y.
RR Lyr	19h 26m	+42°47"	Variable star	7.1–8.1	855 l.y.
M57	18h 54m	+33°02"	Planetary nebula	8.8	2,300 l.y.

VULPECULA AND SAGITTA
THE FOX AND THE ARROW

Tucked alongside the more distinctive constellations of Cygnus and Lyra, these two small and obscure star groups nevertheless contain some interesting objects. The arrow shape of Sagitta, pointed toward the celestial birds Cygnus and Aquila, is the more easily recognized.

Sagitta's brightest star, misleadingly designated Gamma, is one of the few cool M-class stars visible to the naked eye—a dying red giant about 275 light years away. It is also home to Messier 71, a loosely bound globular cluster 13,000 light years from Earth.

The shapeless and faint Vulpecula, meanwhile, supposedly represents a fox running with a goose in its jaws. The goose is supposedly marked by the constellation's brightest star, known as Anser. Other highlights include the Dumbbell Nebula, and the distinctive coathanger-shaped group of stars known as Brocchi's Cluster, or Collinder 399. In 1967, astronomers surveying the constellation with radio telescopes, discovered PSR B1919+21, a steadily flashing radio source that proved to be the first

DUMBBELL NEBULA Also known as Messier 27, the Dumbbell was the first planetary nebula to be discovered, by French astronomer Charles Messier in 17__. 1,350 light years away, it is one of the closest objects of its kind to Earth, and is easy to spot through binoculars as a patch of light one-third the size of the full __

	R.A.	DEC.	TYPE	MAGNITUDE	DISTANCE
ζ Sge	19h 49m	+19°09"	Multiple star	5.5/6.0/8.7	255 l.y.
M27	20h 00m	+22°43"	Planetary nebula	7.4	1,350 l.y.
M71	19h 54m	+18°47"	Globular cluster	6.1	13,000 l.y.

CYGNUS
THE SWAN

Often called the Northern Cross, the bright stars of Cygnus form a prominent pattern in northern skies, encompassing some of the richest starfields in the Milky Way and a variety of fascinating deep-sky objects.

The ancient Greeks viewed Cygnus as one of several mythological swans, flying along the Milky Way. To some, it represented Zeus, the king of the gods himself, taking the form of a swan in order to seduce the beautiful Leda. To others, it was the musician Orpheus, transformed into a swan after his death and placed in the sky alongside his lyre. A third legend saw Cygnus as the loyal friend of Phaeton, a reckless youth who stole the Sun god's chariot and crashed it into the celestial river Eridanus. According to this interpretation, Cygnus repeatedly plunged into the river in a vain attempt to save his friend, and so Zeus preserved him as a swan among the stars.

Whatever its identity, Cygnus is an astronomical treasure trove. Its brightest star, Deneb, is the most intrinsically luminous of all first-magnitude stars, shining with the brilliance of 200,000 Suns and still brilliant across a distance of 3,000 light years.

VEIL NEBULA One of the brightest supernova remnants in the sky, the diaphanous Veil Nebula covers an area of sky roughly six times the diameter of the full Moon, a little way to the southeast of Epsilon Cygni. These shredded sheets of hot gas have now expanded into a shell around 50 light years across. The brightest segments, discovered by William Herschel in 1784, have separate NGC catalog numbers 6960, 6992 and 6995.

	R.A.	DEC.	TYPE	MAGNITUDE	DISTANCE
Albireo (β)	19h 31m	+27°58″	Binary star	3.1/5.1	385 l.y.
μ Cyg	21h 44m	+28°45″	Binary star	4.8/6.2	72 l.y.
61 Cyg	21h 07m	+38°45″	Binary star	5.0	11.4 l.y.
P Cyg	20h 18m	+38°02″	Variable star	c.4.8 (var)	5,500 l.y.
M39	21h 32m	+48°26″	Open cluster	5.5	800 l.y.
NGC 7000	20h 59m	+44°20″	Diffuse nebula	4.0	1,600 l.y.

CYGNUS INSIDE VIEW

CYGNUS X-1

Visible-light observations of this strange object reveal a brilliant blue supergiant, reduced to a feeble 9th magnitude by its distance of more than 8,000 light years. However, X-ray studies reveal something else entirely—a strong source of high-energy X-rays flickering a thousand times per second. The rays do not come directly from the star, but from an otherwise invisible object that orbits it once every 5.6 days and weighs as much as five Suns. This suggests that the secondary object is probably a black hole, pulling gas away from the visible star and forming a superhot disk of X-ray emitting material that spirals inward to its doom.

R.A. 19h 58m, Dec. +35°12˝
Magnitude 8.9
Distance 8,200 light years

CYGNUS A

In visible light, this distant and distorted galaxy can only be detected through large telescopes, but for radio astronomers it is one of the brightest objects in the sky. A pair of tight jets emerge on opposite sides of the visible galaxy and billow out to form a huge double-lobed structure of radio-emitting gas half a million light years across. Cygnus A is one of the nearest and most powerful radio galaxies—a type of active galaxy in which the central supermassive black hole "engine" is hidden from view so that all we see are the effects of material ejected from above and below this central region.

R.A. 19h 59m, Dec. +40°44˝
Magnitude 15.0
Distance 600 million light years

ALBIREO
β CYGNI

Arguably the most beautiful double star in northern skies, Albireo is an observing treat through even the smallest telescope. The system comprises a yellow-orange star of magnitude 3.1 and a contrasting blue-green star of magnitude 5.1. Both stars lie around 385 light years from Earth, but astronomers have not confirmed that they are actually in orbit around each other. In 1976 astronomers confirmed that the brighter component, Albireo A, is itself binary, but its two elements are all but impossible to separate visually, even with professional equipment.

R.A. 19h 31m, Dec. +27°58˝
Magnitude 3.1/5.1
Distance 385 light years

▶ THE WITCH'S BROOM
NGC 6960

The brightest section of the Veil Nebula, marking its western edge, is sometmes known as the Witch's Broom. Here the nebula's light is relatively concentrated, and the naked-eye star 52 Cygni (magnitude 5.3) acts as a useful pointer. This segment of the Veil is roughly 1.5 degrees across—three times the size of the full Moon. At an estimated distance of almost 1,500 light years, meanwhile, the entire Veil complex covers a volume of space 50 light years across.

R.A. 20h 46m, Dec. +30°43˝
Magnitude 7.0
Distance 1,470 light years

CONTINENT IN SPACE

The celebrated North America Nebula covers an area of sky four times the size of the full Moon, east of the constellation's brightest star Deneb. The resemblance to the continent of North America is clear, with a dark dust lane forming the Atlantic coast and Gulf of Mexico. On the other side of this dust lane lies another distinctive shape, the Pelican Nebula IC 5070. Although NGC 7000's "integrated magnitude" is estimated at around 4.0, the size of the nebula means that its light is spread across a large area, so it is hard to pick out from the background Milky Way with the naked eye. Instead, the nebula is best observed under dark, moonless skies with binoculars or a telescope with low magnification and a wide field of view.

IONIZATION FRONT

These colorful peaks are found within the Pelican Nebula, separated from the North America Nebula proper by a canyon of opaque dust but essentially part of the same star-forming complex. The mountainous outcrops have formed as radiation from newborn stars concentrated along the canyon blows away the surrounding gas. Radiation absorbed and then re-emitted by the gas causes it to glow in colors that are linked to its composition—while some gas is ionized (transformed into electrically charged particles), creating the characteristic blue haze above the peaks themselves. Denser areas within the nebula hold out against the torrent of radiation for longer. For instance, the tendril-like structures at the top of each peak form where relatively dense knots of material cast a protective "shadow." These knots may ultimately condense to create further young stars.

▶ MULTIWAVELENGTH MONTAGE

A series of images reveals the inner regions of the North America Nebula in four different wavelengths. At top left, a visible-light view emphasizes the similarity to North America, and also highlights the Pelican Nebula and intervening dust lane. At top right, visible light (blue) has been combined with cooler infrared radiation (red) to show the warm glow of the nebula's dust clouds. Mid-infrared (bottom left) and far-infrared (bottom right) images pierce the dust completely to reveal the glow of newborn stars embedded within the nebula.

R.A. 20h 59m, Dec. +44°20"
Magnitude 4.0
Distance 1,600 light years

ANDROMEDA AND LACERTA
THE PRINCESS AND THE LIZARD

The fairly indistinct branching shape of Andromeda is still easy to locate in the sky because its brightest star is also the northeast corner of the Square of Pegasus. The more compact and distinctive shape of Lacerta can be found to its northwest.

Andromeda is an ancient constellation—in mythology, she was the beautiful daughter of Cepheus and Cassiopeia of Ethiopia. Cassiopeia's boasts about Andromeda enraged the jealous Juno, queen of the gods, who sent the sea monster Cetus to ravage their kingdom until Andromeda was sacrificed. Lacerta is a more recent invention, created by Johannes Hevelius around 1687.

Alpha Andromedae, or Alpheratz, is a slightly variable blue-white star of magnitude 2.1, while Gamma, or Almach, is an attractive multiple. A small telescope will split it into yellow and blue stars of magnitudes 2.3 and 4.8. The blue star has a fainter companion of magnitude 6.1 that is currently drawing away from its closest alignment in 2012, when the pair were indivisible through even the largest telescopes.

BLUE SNOWBALL The planetary nebula NGC 7662, 2,200 light years from Earth, lies midway between Iota and Omicron Andromedae and is one of the easiest objects of its kind for amateurs to detect at magnitude 9.0. A small telescope will show it as a blue-green starlike point, while medium-sized instruments reveal a distinct disk.

	R.A.	DEC.	TYPE	MAGNITUDE	DISTANCE
Alpheratz (α)	00h 08m	+29°05"	Binary star	2.1	97 l.y.
Almach (γ)	02h 04m	+42°20"	Multiple star	2.3/4.8	355 l.y.
M31	00h 43m	+41°16"	Spiral galaxy	3.4	2.5 million l.y.
M32	00h 43m	+40°52"	Elliptical galaxy	8.1	2.5 million l.y.
M110	00h 40m	+41°41"	Elliptical galaxy	8.9	2.7 million l.y.
NGC 7662	23h 26m	+42°33"	Planetary nebula	9.0	2,200 l.y.

ANDROMEDA THE ANDROMEDA GALAXY M31

DUST AND STARS

This unusual view of Andromeda, captured by NASA's Wide-Field Infrared Survey Explorer (WISE) shows the galaxy in a variety of infrared wavelenths. Near-infrared radiation from comparatively hot objects is shown in blue, highlighting the locations of the galaxy's less massive, cooler, and longer-lived stars. Longer wavelengths from cooler regions are shown in green and red—by highlighting regions such as starbirth nebulae, they also trace the locations where hotter, short-lived stars are found. These massive, hot stars only survive for a few million years so they tend to dominate young open clusters and rarely have time to migrate far from the site of their origin. As a result, the warmer yellow areas trace the galaxy's star-forming spiral arms, revealing a complex structure that may have been shaped by past collisions between galaxies.

DOUBLE NUCLEUS

This enhanced image of the galaxy's nucleus from the Hubble Space Telescope reveals two distinct concentrations of stars at the heart of Andromeda. The true center of M31 coincides with the fainter of these two structures, and until recently astronomers had trouble explaining the galaxy's apparent "double nucleus." One popular theory suggested that the brighter nucleus was the remnant of another galaxy absorbed into Andromeda at some point in the past. However, improved computer models suggest that the nucleus of a second galaxy could not persist for long without losing its integrity, so the currently accepted theory is that the brighter concentration comes from a disk of stars orbiting the true center of M31 and bunching up to form a stellar "traffic jam" when they are at their furthest from the nucleus.

▶ **ISLAND UNIVERSE**

The magnificent Andromeda Galaxy is clearly visible with the naked eye under dark skies, as an elongated patch of light about twice the size of the full Moon. Small telescopes will show the bright knot of the galaxy's nucleus floating in the fainter glow of the galaxy's broad disk, but larger instruments or long-exposure photographs are required to bring out features such as the dark dust lanes that help define the spiral structure. With a diameter of around 200,000 light years, M31 is far larger than our galaxy but seems to have less mass. Despite this, it forms the other major gravitational center in our "Local Group" of galaxies, and is powerful enough to attract satellite galaxies including M32 and M110, shown in this image.

R.A. 00h 43m, Dec. +41°16″
Magnitude 3.4
Distance 2.5 million light years

PERSEUS

PERSEUS

This constellation represents a hero whose legend encompasses much of the northern sky, but its awkward shape and position against rich Milky Way starfields can still make it challenging to identify.

In Greek myth, Perseus was an exiled prince of Argos. With the help of the goddess Athena, he slew the gorgon Medusa, then went on to rescue Andromeda from the monster Cetus. He is usually depicted brandishing Medusa's head in one hand.

The constellation's most famous star, Beta Persei, or Algol, is said to mark the eye of Medusa's head brandished in the hero's hand. Algol normally shines at a steady magnitude 2.1, but dips abruptly to magnitude 3.4 for ten hours in a cycle that repeats every 2 days and 21 hours. It is the most prominent "eclipsing binary" in the sky—a system in which two stars in close orbit around each other occasionally pass in front of each other, reducing the system's overall brightness. Algol's behavior has been recognized since at least 1670, but its name, from the Arabic for "demon," suggests its strange nature may have been known since ancient times.

CALIFORNIA NEBULA This softly glowing nebula, catalogued as NGC 1499, lies close to the magnitude 4.0 star Xi Persei and is roughly five times the width of the full Moon. It takes its name from its resemblance to the U.S. state in long-exposure photographs. Xi Persei, at the center of this image, is one of the most massive, hottest and most luminous stars in the sky, with the mass of 40 Suns and a total luminosity 330,000 times that of our star. It lies around 1,800 light years from Earth.

	R.A.	DEC.	TYPE	MAGNITUDE	DISTANCE
Algol (β)	03h 08m	+40°57"	Binary star	2.1–3.4 (var)	93 l.y.
η Per	02h 51m	+55°54"	Visual double	3.8, 8.4	c.1300 l.y.
M34	02h 42m	+42°47"	Open cluster	5.5	1,400 l.y.
M76	01h 42m	+51°34"	Planetary nebula	10.1	8,200 l.y.
NGC 869/884	02h 21m	+57°08"	Open clusters	4.3, 4.4	7,100, 7,400 l.y.
Melotte 20	03h 27m	+49°07"	Open cluster	1.2	600 l.y.

DOUBLE CLUSTER
NGC 869 AND 884

Midway between Perseus and Cassiopeia lies one of the most beautiful sights in northern skies. What looks to the naked eye like a pair of fuzzy "stars," transforms in binoculars or a small telescope into a pair of rich open clusters studded with stars. The two clusters are separated by around 300 light years and so are not bound together by gravity, but they share a common origin in a group of stars and clusters known as the Perseus OB1 Association. Both clusters are relatively young—NGC 869 is around 19 million years old, and NGC 884 is 12.5 million years old—so they are dominated by a mix of massive, brilliant blue-white stars and even more massive red and orange giants that are already approaching the end of their lives.

R.A. 02h 21m, Dec. +57°08"
Magnitude 4.3 and 4.4
Distance 7,100 and 7,400 light years

PERSEUS CLUSTER
ABELL 426

A little way to the east of Algol lies the heart of one of the most massive galaxy clusters in the local Universe, around 240 million light years from Earth. The Perseus Cluster dwarfs even the nearby Virgo Cluster (see page 102), containing as it does thousands of individual galaxies. The space around the cluster galaxies is filled with multi-million-degree, X-ray-emitting gas that has been stripped from the galaxies themselves during billions of years of collisions and close encounters. The giant galaxy NGC 1275 is embedded in the hottest part of this gas at the heart of the cluster.

R.A. 03h 18m, Dec. +41°30"
Magnitude 12.6
Distance 240 million light years

▶ PERSEUS A

The giant galaxy at the center of
the Perseus Cluster, NGC 1275,
coincides with a strong radio source
known as Perseus A. Hubble Space
Telescope images have revealed the
galaxy's peculiar structure—most
likely a result of a collision between
two spiral galaxies. The collision
has triggered a wave of activity in
the nucleus of NGC 1275, pouring
material into the supermassive
black hole at its core, and triggering
the emission of intense light and
other radiations including radio
waves. According to one popular
theory, mergers between spirals
within crowded galaxy clusters
strip them of star-forming gas and
eventually transform them into
giant elliptical galaxies.

R.A. 03h 19m, Dec. +41°30″
Magnitude 12.6
Distance 237 million light years

PISCES
THE FISHES

This relatively faint but distinctive zodiac constellation consists of two chains of stars lying to the east of the brighter stars of the Great Square of Pegasus. Each chain represents a fish, and they are often depicted tied together at their tails, with Alpha Piscium, or Alrescha, marking the cord that binds them.

In ancient times, however, the fish were visualized as swimming freely, and were thought to represent Aphrodite and Cupid fleeing the approach of the sea monster Typhon (nearby Cetus). Alrescha is a good double star through a moderate telescope, while Zeta Piscium is another double that can be resolved into its components through even the smallest instrument.

The "First Point of Aries," marking the northern vernal equinox where the ecliptic crosses from the southern to the northern hemisphere of the sky (see page 16) lies near the constellation's southern edge. It marks the origin point for the entire equatorial coordinate system that astronomers use for mapping the sky.

MESSIER 74 A little to the east of Eta Piscium, this spectacular spiral galaxy is best seen through larger telescopes at low magnification. Although relatively nearby at 30 million light years, it lies face-on to Earth, so its light is spread across the sky. Long-exposure photographs, however, reveal a beautiful geometric structure based around two spiral arms. Messier 74 is often cited as a "Grand Design" spiral, in which the spiral arms, sites of new star formation, are very well defined.

	R.A.	DEC.	TYPE	MAGNITUDE	DISTANCE
Alrescha (α)	02h 02m	+02°46"	Multiple star	4.2/5.2	140 l.y.
ζ Psc	01h 14m	+07°35"	Multiple star	5.2/6.4	148 l.y.
ρ Psc	01h 26m	+19°10"	Visual double	5.3, 5.5	85, 280 l.y.
M74	01h 37m	+15°47"	Spiral galaxy	10.0	30 million l.y.

ARIES AND TRIANGULUM

THE RAM AND THE TRIANGLE

The relatively faint zodiac constellation of the Ram is somewhat lacking in interesting objects, but a nearby compact triangle of faint stars offers the wonders of one of our closest galactic neighbors.

The hunched pattern of Aries has been identified as a crouching ram since well before the heyday of Greek astronomy, but today the constellation is usually seen as the ram that bore the golden fleece sought by Jason and the Argonauts. The wedge-shaped triangle of stars to its north was first noted by Greek astronomers on account of its similarity to the capital letter Delta (Δ).

Alpha Arietis, or Hamal, is an orange giant of magnitude 2.0, about 66 light years from Earth. It is a rare example of a star whose size has been directly measured, and is known to be 14.7 times bigger than the Sun. Gamma Arietis or Mesarthim, meanwhile, is an attractive multiple, 205 light years away—its twin white stars of magnitudes 4.6 and 4.7 are easily split by the smallest telescope, while the sharp-eyed may spot a smaller orange star of magnitude 9.6 that orbits around them at a greater distance.

GALACTIC ENCOUNTER Northwest of Beta Arietis lies a pair of galaxies best seen through a medium-sized telescope. NGC 678 (at left in this photograph) is an edge-on spiral with a dark lane of dust across its center. NGC 680, meanwhile, is an elliptical—a ball of stars lacking the gas to form new stars. Both galaxies are around 120 million light years away, and NGC 680 is being distorted by its neighbor's gravity.

	R.A.	DEC.	TYPE	MAGNITUDE	DISTANCE
Mesarthim (g Ari)	01h 54m	+19°18"	Multiple star	4.6/4.7	205 l.y.
λ Ari	01h 58m	+23°36"	Binary star	5.0/7.7	130 l.y.
M33	01h 34m	+30°39"	Spiral galaxy	5.7	2.7 million l.y.

TRIANGULUM TRIANGULUM GALAXY M33

TRIANGULUM GALAXY
MESSIER 33

Triangulum's greatest showpiece lies close to its western edge. Messier 33 is the closest major galaxy after the Andromeda spiral, Messier 31, and unlike M31 it presents itself face-on to Earth so that its entire spiral structure is displayed. Unfortunately, this means that the galaxy's light is spread out across an area of sky larger than the full Moon, making it surprisingly hard to see. It is best located in very dark skies using binoculars or a telescope on very low power to spot a contrasting patch of light against the dark sky. Triangulum's disappointingly faint appearance is not entirely due to our viewing angle—it really is a far less impressive galaxy than either M31 or the Milky Way. With a much fainter and looser spiral structure than its near neighbors in the Local Group, M33 is classed as a "flocculent" spiral—only telescopes of moderate size will be able to pick out any detail within it.

R.A. 01h 34m, Dec. +30°39"
Magnitude 5.7
Distance 2.7 million light years

INFRARED VIEW

This view of Triangulum from NASA's Spitzer Space Telescope reveals a range of infrared (heat) radiation from objects too faint to shine in visible light. Red indicates the longest-wavelength, coolest radiations, while green and blue indicate successively shorter wavelengths and warmer material. Stars (including some foreground stars within our own galaxy) appear blue, while colder interstellar dust within the Triangulum Galaxy glows red. While the central blue haze coincides with the extent of the visible galaxy, tendrils of cooler red gas and dust extend well beyond M33's visible limits. This invisible material seems to have migrated outward from the galaxy's inner regions, but astronomers are still puzzling over exactly how this could have happened.

► **FOUNTAIN OF STARBIRTH**
NGC 604

While the Triangulum Galaxy lacks any regimented structure in its star-forming regions, it is still home to some enormous starbirth nebulae. These include NGC 604, one of the largest nebulae in the Local Group. Shown here in a Hubble Space Telescope image, it has an estimated diameter of 1,500 light years. Its diameter is around 40 times that of Orion's Great Nebula, and it is thought to be more than 6,000 times as luminous as that famous starbirth region.

R.A. 01h 34m, Dec. +30º47¨
Magnitude 14.0
Distance 2.7 million light years

TAURUS
THE BULL

One of the most distinctive constellations in the sky, Taurus represents the figure of a celestial bull charging toward the hunter Orion. Rich in interesting stars and deep-sky objects, it is also a zodiac constellation that plays host to the Moon and planets.

Lying a little way to the north of the celestial equator, Taurus is visible from nearly all inhabited parts of the world, and is a familiar fixture of northern hemisphere skies on fall and winter nights. It is one of the few constellations that actually resembles the figure after which it is named—the forequarters of a charging bull—and has been recognized by stargazers since prehistoric times.

The distinctive V-shape of the Hyades star cluster represents the face of the bull, with brilliant red Aldebaran, a foreground star, marking its eye. Meanwhile, two bright stars (one of which, Beta Tauri or Alnath, is shared with neighboring Auriga) indicate the tips of its horns, fainter chains of stars mark the beast's forelegs, and the Pleiades star cluster indicates the location of its shoulders.

FACE OF THE BULL The V-shaped Hyades is the most prominent star cluster in the entire sky—close enough for individual members to be easily distinguished, but far enough away to form a distinct and compact grouping. Evidence from an array of measurements suggests the cluster is around 150 light years from Earth, although the brilliant orange foreground star Aldebaran is at only half this distance. The cluster contains several hundred stars that formed around 600 million years ago.

	R.A.	DEC.	TYPE	MAGNITUDE	DISTANCE
Aldebaran (α)	04h 36m	+16°31″	Variable star	0.8–1.0 (var)	65 l.y.
χ Tau	04h 22m	+25°38″	Binary star	5.4/7.6	270 l.y.
Hyades	04h 27m	+16°	Open cluster	N/A	150 l.y.
M1	05h 35m	+22°01″	Supernova remnant	8.4	6,500 l.y.
M45	03h 47m	+24°	Open cluster	1.6	440 l.y.

◄ SEVEN SISTERS

Named for the daughters of Atlas and Pleione in Greek mythology, the Pleiades form an obvious hook-like shape on the shoulder of Taurus. Although clearly visible with the naked eye, the group is not quite as bright as their integrated magnitude of 1.6 would suggest. Observers with average eyesight can usually count six of the "Seven Sisters," but those with keen eyes can spot seven or even more. Binoculars or a small telescope with low magnification can reveal the cluster in all its glory—it is thought to contain at least 1,000 members within a region around 100 light years across.

R.A. 03h 47m, Dec. +24°07"
Magnitude 1.6
Distance 440 light years

INFRARED PLEIADES

This unusually colorful view of the Pleiades was captured using NASA's infrared Spitzer Space Telescope, and traces density variations in the veil of gas that surrounds the Pleiades Cluster. Red colors highlight the densest regions of the gas cloud, while yellows and greens mark its tenuous outlying regions. This image shows the regions around three of the bright "sister" stars—Alcyone (center), right, embedded in the densest area of nebulosity). Despite appearances, the material around the Pleiades seems not to be linked to their origin—instead it is a separate region of interstellar gas, through which the young cluster just happens to be drifting.

MEROPE NEBULA
NGC 1435

Also known as Tempel's Nebula, NGC 1435 marks the densest concentration of gas and dust in the vicinity of the Pleiades. Shining bluish-white thanks to light scattered from nearby Merope, it forms a patch of light roughly the size of the full Moon. However, thanks to its faint apparent magnitude of around 13.0, it is only visible through large telescopes or in long photographic exposures. The Hubble Space Telescope image above captures the ghostly appearance of the brightest knot within the nebula, a region given its own classification as IC 349.

R.A. 03h 46m, Dec. +23°54"
Magnitude 13.0
Distance 440 light years

TAURUS THE CRAB NEBULA M1

CRAB PULSAR
PSR B0531+21

At the heart of the Crab Nebula lies a rapidly spinning neutron star. This is the surviving core of the massive star whose explosion was seen in AD 1054, compressed to the size of a city. Because the collapsed core retains much of the angular momentum and magnetic field of its parent star, it spins at a tremendous rate and channels most of its radiation into two narrow beams aligned with its magnetic poles. As these beams sweep around the sky like a cosmic lighthouse, they are briefly pointed toward Earth, creating a flashing beacon of radio waves known as a pulsar. The Crab Pulsar has now been detected in visible light and X-rays, as shown in this composite image from the Hubble Space Telescope and Chandra X-Ray Observatory.

X-RAY VIEW

An image from NASA's Chandra X-Ray Observatory maps high-energy radiation from the region around the Crab Pulsar. Two jets of X-ray-emitting particles can be seen clearly emerging from above and below the pulsar itself, while a series of concentric waves ripple out into a general X-ray-emitting haze around the collapsed star. The X-rays are a form of synchrotron radiation, emitted by electrons travelling at high speed, and this emission cloud is thought to play a crucial role in transferring energy from the pulsar itself into the nebula. Particles falling onto the rapidly spinning stellar remnant are accelerated to high speeds and ricochet back out to energize the surrounding gas clouds.

▶ **STELLAR REMNANT**

At magnitude 8.4, the Crab Nebula is bright enough to be spotted through binoculars or a small telescope as a fuzzy patch of light northwest of Zeta Tauri (the tip of Taurus's southern horn). Larger instruments or long-exposure images transform it into a shredded web of glowing gas some 11 light years across—the most celebrated supernova remnant in the sky. The Crab was first recorded by British astronomer John Bevis in 1734, and a few decades later French comet hunter Charles Messier listed it as number 1 in his catalog of objects that might be mistaken for comets. It was not until 1939 that astronomers first associated it with the enormous stellar outburst recored by stargazers around the world in AD 1054.

R.A. 05h 35m, Dec. +22°01′
Magnitude 8.4
Distance 6,500 light years

GEMINI
THE TWINS

The bright stars Castor and Pollux draw attention to this constellation—small wonder that it has been associated with twins or pairs by cultures across the world.

Ancient Chinese astronomers saw these two stars as the balanced principles of yin and yang, and the Romans associated them with Romulus and Remus, founders of Rome itself. However, their common names represent the twin sons of the Spartan Queen Leda, mortal Castor and immortal Pollux, who joined Jason's crew in the quest for the golden fleece.

Fittingly, Castor is a celebrated multiple star, about 52 light years from Earth, with a combined naked-eye magnitude of 1.6. Small telescopes can split it to reveal a pair of blue-white stars of magnitudes 1.9 and 2.9, and reveal a third, more distant red dwarf of magnitude 9.3. Each of these stars is a double in its own right, although the components cannot be separated visually. Despite its designation as Beta Geminorum, Pollux is actually the brighter of the twin stars, shining at magnitude 1.2. In contrast to Castor, it is a stellar singleton—an orange giant around 34 light years away.

ESKIMO NEBULA This complex planetary nebula, also known as NGC 2392, is visible through a small telescope to the southeast of Delta Geminorum. It lies 3,000 light years away and shines at magnitude 10.1. Despite its disk-like appearance, the nebula actually consists of two lobes of material being blown away from the central dying star—one directly toward us, and the other in the opposite direction.

	R.A.	DEC.	TYPE	MAGNITUDE	DISTANCE
Castor (α)	07h 35m	+31°53"	Multiple star	1.9/2.9	52 l.y.
Pollux (β)	07h 45m	+28°02"	Planetary system	1.2	34 l.y.
ζ Gem	07h 04m	+20°34"	Variable star	3.7–4.2 (var)	1,170 l.y.
μ Gem	06h 23m	+22°31"	Variable star	2.7–3.0 (var)	230 l.y.
M35	06h 09m	+24°20"	Open cluster	5.3	2,800 l.y.
NGC 2392	07h 29m	+20°55"	Planetary nebula	10.1	3,000 l.y.

CANCER

THE CRAB

The faintest of the zodiac constellations, Cancer was known in medieval times as the "Dark Sign" on account of its obscurity. It is best found by looking between the brighter stars of Leo and Gemini.

Despite its faintness, Cancer has been associated with a crab since ancient times. Its origins are untraceable, but the Greeks saw it as a crab that was crushed beneath the foot of Hercules as he fought the serpent Draco. The Egyptians, however, saw the constellation as a sacred scarab beetle.

Despite its designation, Alpha Cancri or Acubens is actually the constellation's fourth-brightest star at magnitude 4.3. A moderate telescope reveals that this white star, 174 light years from Earth, has a faint companion of magnitude 11.9. Zeta Cancri, meanwhile, is a quadruple system—small telescopes show two components of magnitudes 5.1 and 6.2, moderate instruments reveal that the brighter of these is itself an evenly matched double and professional instruments show that the fainter star has a dim red-dwarf companion.

BEEHIVE CLUSTER Cancer's great claim to fame is the star cluster Messier 44, known as the Beehive or Praesepe (the manger). Composed of around 200 stars spread across an area roughly three times the diameter of the full Moon, the Beehive lies around 580 light years from Earth, and is clearly visible to the naked eye. The great Italian astronomer Galileo Galilei was the first person to resolve its individual stars in the early 1600s, using one of his early, self-built telescopes.

	R.A.	DEC.	TYPE	MAGNITUDE	DISTANCE
Acubens (α)	08h 58m	+11°51"	Multiple star	4.3	174 l.y.
ζ Cnc	08h 12m	+17°38"	Multiple star	5.1/6.2	83 l.y.
55 Cnc	08h 53m	+28°20"	Planetary system	6.0	41 l.y.
M44	08h 40m	+19°59"	Open cluster	3.7	580 l.y.
M67	08h 50m	+11°49"	Open cluster	6.1	2,700 l.y.

LEO AND LEO MINOR

THE LION AND THE LESSER LION

The zodiac constellation of the Lion is one of the most instantly identifiable in the sky, and a rare case of a constellation that really looks like the object it claims to represent. Sadly, the same cannot be said for its smaller neighbor.

Leo has been recognized as the figure of a lion by almost every culture (though Chinese astronomers saw it as a horse). Since classical times, it has been associated with the Nemaean lion fought by the hero Hercules. A curve of stars marking its forequarters is widely known as the Sickle. At its base lies Leo's brightest star, Regulus, whose name means "little king." This brilliant white star of magnitude 1.35 lies about 77 light years from Earth. Small telescopes reveal the brighter of two companions that orbit it at a considerable distance. Regulus's location, almost exactly on the ecliptic, allows the Moon and planets to occasionally pass directly in front of it, creating a beautiful event known as an occultation.

In contrast to Leo, faint and shapeless Leo Minor is a late addition to the skies, invented by Polish astronomer, Johannes Hevelius in the 17th century

HICKSON 44 Located in the neck of the lion between Gamma and Zeta Leonis, Hickson Compact Group 44 is a small cluster of four galaxies, bound together in a gravitational waltz around 60 million light years from Earth. Its brightest members, the edge-on spiral NGC 3190 and the elliptical NGC 3193 (both in the upper left corner) can be spotted with a small amateur telescope.

	R.A.	DEC.	TYPE	MAGNITUDE	DISTANCE
Regulus (α)	10h 08m	+11°58″	Multiple star	1.35/8.1/13.5	77 l.y.
Algieba (γ)	10h 20m	+19°50″	Binary star	2.0	126 l.y.
R Leo	09h 48m	+11°26″	Variable star	4.3–11.6 (var)	390 l.y.
M65	11h 19m	+13°06″	Spiral galaxy	10.3	34 million l.y.
M66	11h 20m	+13°00″	Spiral galaxy	8.9	35 million l.y.
M95	10h 44m	+11°42″	Barred spiral gal.	11.4	38 million l.y.
M96	10h 47m	+11°49″	Spiral galaxy	10.1	32 million l.y.

BARRED SPIRAL
MESSIER 95

This small but exquisite spiral galaxy is one member of the "Leo I" group that lies to the south of the lion's body, about halfway along its length. It is a particularly extreme example of a barred spiral with a "circumnuclear ring"—a circular zone of star formation roughly 2,000 light years across that runs all around the nucleus connecting the ends of the central bar and extending out into the galaxy's spiral arms. Our own Milky Way galaxy, recently revealed as a barred spiral itself, is thought to have a similar circumnuclear ring.

IMPERFECT SPIRAL
MESSIER 96

The largest and brightest galaxy in the Leo I group, M96 is still beyond the reach of binoculars, but just about detectable through a small telescope. It forms a stark contrast to the perfection of its neighbor M95, with warped and faint spiral arms, a nucleus displaced from the center and an asymmetric distribution of gas and dust. One of the galaxy's most enticing features, revealed in this photograph from the European Southern Observatory's Very Large Telescope, is the fact that it lies in front of a more distant galaxy cluster, including a beautiful edge-on spiral at upper left.

GALAXY ON EDGE
NGC 3628

This unbarred spiral galaxy, seen almost edge-on from Earth, is a member of the Leo Triplet galaxy group along with Messier 65 and M66. While smaller telescopes can show it as an elongated smudge of light, larger instruments reveal a dark lane of dust, associated with the outer edge of the spiral arms, that runs down the length of the galaxy. Although unseen in this image, a long "tidal tail" extends from one side of the galaxy, reaching more than 300,000 light years into intergalactic space.

► WARPED BEAUTY
MESSIER 66

Leo's most spectacular galaxy is Messier 66, the largest and brightest member of the so-called "Leo Triplet" group that nestles below Theta Leonis in the lion's hindlimb. With a diameter of around 95,000 light years, it is roughly the same size as the Milky Way, but close encounters with its galactic neighbors in the past have warped its disk and spiral arms. They have also concentrated much of the galaxy's mass around its nucleus, while apparently pulling the nucleus away from the geometric center of the spiral.

R.A. 10h 44m, Dec. +11°42"
Magnitude 11.4
Distance 38 million light years

R.A. 10h 47m, Dec. +11°49"
Magnitude 10.1
Distance 32 million light years

R.A. 11h 20m, Dec. +13°35"
Magnitude 9.4
Distance 35 million light years

R.A. 11h 20m, Dec. +13°00"
Magnitude 8.9
Distance 35 million light years

COMA
BERENICES
QUEEN BERENICE'S HAIR

Though faint and unprepossessing to the naked eye, this constellation reveals an unexpected beauty in dark skies or through binoculars. It is home to one of the closest star clusters to Earth.

Coma, as it is usually known for short, is a rare example of a constellation named to honor a historical, rather than mythological, figure. Queen Berenice was the wife of Ptolemy III, ruler of Egypt in the third century BC, who offered to cut off her luxurious tresses in return for her husband's safety

UNEXPECTED SPIRAL
NGC 4911

This beautiful Hubble Space Telescope image shows a large, face-on spiral galaxy near the center of the Coma Cluster about 320 million light years from Earth. NGC 4911 is something of a mystery since it is clearly a site of considerable star-forming activity, revealed by the intense pink nebulae that mark out its spiral arms. Such activity is unusual at the center of a major galaxy cluster, where intergalactic collisions and close encounters tend to strip galaxies of their star-forming materials.

NEEDLE GALAXY
NGC 4565

This immaculate edge-on galaxy lies embedded in the midst of the Coma star cluster Melotte 111, but seems to be independent of both the Virgo and Coma galaxy clusters. Discovered by William Herschel in 1785, it is thought to be a barred spiral, although the bar is hidden from our direct view. At magnitude 10.4, it can be located with a small telescope, while a medium-sized instrument will show the dark dust lane running along its length and the central bulge of the nucleus.

RING OF STARBIRTH
NGC 4314

This unusual barred spiral galaxy, roughly 40 million light years from Earth, has recently undergone a remarkable burst of star formation. This can be traced through the bright ring around the core, shown in blue and purple in this Hubble Space Telescope image. The ring, a thousand light years across, is highlighted by the intense blue of young star clusters, and the purple glow of star-forming hydrogen gas. The color of the star clusters suggests they are extremely young and that the entire starburst has taken place over the last few million years.

▶ COMA CLUSTER

While southern Coma Berenices plays host to many galaxies that properly belong in the Virgo Cluster, the northern part of the constellation is also home to Coma's own galaxy cluster. The Coma Cluster is far more distant, at around 320 million light years from Earth. Its brightest galaxies are the NGC 4874 and 4889, both giant ellipticals formed from the merger and absorption of numerous smaller galaxies in the cluster's packed central regions. This false-color view combines a visible-light image in blue with infrared data in red and green.

R.A. 13h 00m, Dec. +27º47"
Magnitude 12.8
Distance 320 million light years

R.A. 12h 36m, Dec. +25º59"
Magnitude 10.4
Distance 40 million light years

R.A. 12h 22m, Dec. +29º53"
Magnitude 11.4
Distance 40 million light years

R.A. 13h 00m, Dec. +27º58"
Magnitude 11.4
Distance 320 million light years

VIRGO
THE MAIDEN

The second-largest constellation in the sky, Virgo is highlighted by the presence of the brilliant star Spica. However, its greatest treasures are telescopic, for Virgo plays host to a wealth of relatively nearby galaxies.

Virgo has been associated with a fertility or harvest goddess since ancient times. The first Babylonian astronomers already associated these stars with the image of a maiden holding an ear of grain or corn (represented by Spica). The Greeks saw her as Persephone, daughter of the harvest goddess Ceres, but also interpreted her as Dike, the goddess of justice holding the scales (Libra) in her raised hand.

Spica lies very close to the ecliptic and is a multiple star system. Its brightest elements, shining at a combined magnitude 1.0, are twin, hot blue-white stars 260 light years away, orbiting each other in just four days and far too tightly bound to be separated by even the largest telescopes. The twin yellow-white stars of Gamma Virginis, or Porrima, on the other hand, can be separated even with small telescopes.

GALAXY CLUSTER Northern Virgo is home to the closest major galaxy cluster to Earth, centered around 54 million light years away. Galaxies are gregarious by nature, and the Virgo cluster contains several dozen major spirals and ellipticals, centered around the giant elliptical Messier 87, shown here.

	R.A.	DEC.	TYPE	MAGNITUDE	DISTANCE
Spica (α)	13h 25m	-11°10″	Multiple star	1.0	260 l.y.
Porrima (γ)	12h 42m	-01°27″	Binary star	2.9	32 l.y.
M49	12h 30m	+08°00″	Elliptical gal.	8.4	60 million l.y.
M87	12h 31m	+12°24″	Elliptical gal.	8.6	60 million l.y.
M90	12h 37m	+13°10″	Spiral galaxy	10.3	60 million l.y.
M104	12h 40m	-11°37″	Spiral galaxy	9.0	28 million l.y.

MESSIER 87

The Virgo Cluster's central giant elliptical galaxy is the closest of its kind to Earth. Through small telescopes it looks somewhat similar to a globular cluster, but in reality it is one of the largest known galaxies. A million million stars jostle in a core region roughly 160,000 light years in diameter, while many more orbit in a halo that stretches out to 250,000 light years from the center. Radio signals and a visible jet of material shooting from the core show that M87 has an active nucleus—a supermassive black hole feeding on material from its surroundings. Hubble Space Telescope images have revealed the dusty skeleton of a cannibalized galaxy embedded within M87, and this material is probably responsible for the ongoing activity.

R.A. 12h 31m, Dec. +12°24˝
Magnitude 8.6
Distance 60 million light years

MESSIER 58

This barred spiral galaxy, 68 million light years from Earth, lies on the far side of the Virgo Cluster, but is one of its brightest galaxies. Small telescopes can locate its bright nucleus, but may have difficulty distinguishing the galaxy from nearby ellipticals—larger instruments are needed to pick up the light of the spiral arms. This infrared image from the Spitzer Space Telescope distinguishes between relatively sedate, mature stars spread throughout the galaxy's central core, bar and disk (shown in blue), and gas and dust in star-forming regions concentrated in the spiral arms (shown in red).

R.A. 12h 38m, Dec. +11°49˝
Magnitude 9.7
Distance 68 million light years

► **CENTRAL REGIONS**

This mosaic view covers a broad span of the Virgo Cluster with an area equivalent to 12 full Moons, revealing two clear concentrations of galaxies within. At lower left, dozens of galaxies cluster around the giant elliptical M87, while at upper right, they gather around two smaller ellipticals, M86 and M84. A third concentration, not shown in this image, is focused on another large elliptical called M49. The three groups are in the process of merging together into a single giant cluster—a process that all clusters pass through and which can help astronomers estimate their age.

R.A. 12h 30m, Dec. +08°00'
Magnitude 9.4
Distance 68 million light years

VIRGO THE SOMBRERO GALAXY M104

EMBEDDED DISK

This spectacular galaxy, unrelated to the Virgo Cluster, lies near the constellation's southern edge around 28 million light years from Earth. It appears to be an elliptical ball of stars with a bright disk within it, as highlighted by this composite of Spitzer Space Telescope infrared data and Hubble Space Telescope images. The ring contains a mix of cool dust (highlighted in red), but also the majority of cold atomic and molecular hydrogen gas—the raw materials that fuel ongoing star formation. In contrast, the galaxy's central regions are lacking in these materials and therefore lack the signs of ongoing star formation.

X-RAY VIEW

This composite adds high-energy X-ray data from NASA's Chandra X-Ray Observatory to a visible-light view from Hubble. X-ray emissions are associated with violent processes and extreme objects—typically the environments around black holes. The Chandra observations show a number of X-ray sources scattered through the Sombrero's elliptical halo, and a blaze of radiation from the region around the nucleus, where a supermassive black hole is thought to lurk. This active nucleus is subdued compared to more violent objects such as quasars and Seyfert galaxies, so the black hole is probably feeding only slowly on gas that drifts within its grasp.

▶ SILHOUETTED RING

The Sombrero's outer dust lane is silhouetted starkly against the galaxy's brilliant core in this stunning Hubble Space Telescope image. The ring has an estimated diameter of around 50,000 light years, making it roughly half the size of our own galaxy's main disk. Nevertheless, at a distance of 28 million light years, the Sombrero appears as one of the brightest galaxies in Earth's skies. It is visible through a small telescope or good binoculars, but larger instruments or long-exposure photographs are needed to identify features such as the central bulge or the dust lane.

LIBRA
THE SCALES

This faint constellation lies midway between the bright stars Spica in Virgo and Antares in Scorpius. As the only sign of the zodiac not to represent a living creature, it has often been associated with its brighter neighbors.

In classical times, Libra did not even exist—it was seen as Chelae Scorpionis, the extended claws of neighboring Scorpio. Libra was established as an independent constellation around the first century AD, by which time it had switched its allegiance to Virgo. It is now often seen as the scales of justice, upheld by Virgo in her guise of the goddess Iusticia.

Zubenelgenubi (Alpha Librae) is a wide double star, easily separated through binoculars into individual components of magnitudes 2.2 and 5.2. These two stars are 77 light years from Earth, orbiting each other in around 200,000 years. The bright primary star is a binary in its own right, although it is inseparable through even the largest telescopes. Zubeneschamali (Beta Librae) meanwhile, is one of only a very few stars in the sky to display a noticeably greenish color.

GLIESE 581 This red dwarf to the northeast of Beta Librae lies just 22 light years from Earth and is home to one of the closest known planetary systems. A Neptune-sized planet was discovered here in 2005, and since then at least two, and possibly as many as five more, have been identified. These include Gliese 581c, one of the most Earth-like exoplanets known, and Gliese 581d, which orbits in its solar system's "habitable zone" where liquid water can persist on its surface.

	R.A.	DEC.	TYPE	MAGNITUDE	DISTANCE
Zubenelgenubi (α)	14h 51m	-16°00"	Multiple star	2.2/5.2	77 l.y.
ι Lib	15h 12m	-19°47"	Multiple star	4.5	142 l.y.
48 Lib	15h 58m	-14°17"	Variable star	c.4.9 (var)	510 l.y.
NGC 5897	15h 17m	-21°01"	Globular cluster	8.5	40,000 l.y.

SERPENS

THE SNAKE

Uniquely in the sky, this constellation is split into two distinct parts—Serpens Caput (the snake's head) and Serpens Cauda (the snake's tail)—on either side of the larger constellation Ophiuchus.

This region of the sky has been interpreted as a figure wrestling a dragon since the earliest times. Ancient Greek astronomers associated it with a more benign snake wielded by Asclepius, god of medicine. Serpens Cauda is home to a rich region of the Milky Way including the spectacular Eagle Nebula and its associated star cluster Messier 16, home to the celebrated "Pillars of Creation." Serpens Caput, meanwhile, looks toward a region of space above the plane of the Milky Way, and is home to the bright globular cluster M5.

Alpha Serpentis, or Unukalhai, (the "serpent's neck"), is an orange giant 73 light years from Earth, shining at magnitude 2.6. Delta Serpentis is a multiple-star system consisting of a brighter pair (magnitudes 4.2 and 5.2) 210 light years away, orbited by a much fainter pairing of red dwarfs visible only through larger telescopes.

SERPENS SOUTH Dense clouds of dust obscure parts of the Milky Way in southern Serpens Cauda, at the base of the long Cygnus Rift (see page 65). In 2007, scientists used the infrared vision of NASA's Spitzer Space Telescope to peer through the dust and discover a young star cloud beyond. The Serpens South Cluster consists of around 50 stars crammed into a region roughly five light years across, some 850 light years from Earth, including many that are still in the process of formation.

	R.A.	DEC.	TYPE	MAGNITUDE	DISTANCE
θ Ser	18h 56m	+04°12"	Multiple star	4.6/5.0/6.7	130 l.y.
ν Ser	17h 21m	-12°51"	Binary star	4.3/8.4	193 l.y.
M5	15h 19m	+02°05"	Globular cluster	5.6	24,500 l.y.
M16	18h 19m	-13°47"	Open cluster	6.4	7,000 l.y.

SERPENS INSIDE VIEW

HOAG'S OBJECT

When U.S. astronomer Art Hoag discovered this beautiful wheel in space in northwestern Serpens Caput in 1950, he was unsure whether it was a nearby planetary nebula or an unusual distant galaxy. Later observations confirmed that Hoag's Object is in fact a rare "ring galaxy" 800 million light years from Earth. Other ring galaxies have since been identified, but they usually have features that can be traced back to a collision with a smaller galaxy—the shock wave triggers a ring of star formation. Hoag's Object, however, seems to have formed without such interactions—astronomers are still unsure of its origin.

MESSIER 5

Just visible to the naked eye in dark skies, M5 is the brightest of several globular clusters scattered across Serpens and neighboring Ophiuchus. It appears as an impressive fuzzy "star" through binoculars and a small telescope will begin to resolve individual stars in its outer reaches. The cluster is about 165 light years across and contains several hundred thousand stars—perhaps as many as half a million. It is also one of the oldest known globular clusters at 13 billion years old, and is particularly rich in variable stars of the "RR Lyrae" type, which have been used to accurately measure its distance.

ARP 220

This unusual object in Serpens is known by its catalog entry in the *Atlas of Peculiar Galaxies* published by U.S. astronomer Halton Arp in 1966. Arp identified 338 galaxies whose structure did not conform to normal classifications such as spiral, elliptical and irregular. In the ensuing decades, improved observations have shown that many of his peculiar objects are in fact pairs of colliding galaxies, and satellite observatories have shown other unusual features. Arp 220, for example, emits extremely large amounts of infrared (heat) radiation, thought to be generated by massive bursts of star formation as the galaxies collide and merge.

▶ SEYFERT'S SEXTET
NGC 6027

Discovered by U.S. astronomer Carl Seyfert in the late 1940s, this compact galaxy group in Serpens Caput was at first thought to contain six galaxies. However, later studies showed that the small face-on spiral is a far more distant background object, and the diffuse, fuzzy object at lower right is a "tidal tail" of material pulled away from one of the other interacting galaxies. Nevertheless, the remnant cluster is still impressive, with four galaxies (three edge-on spirals and one elliptical in the center with a prominent dust lane) jammed together in a region of space just 100,000 light years across.

SERPENS THE EAGLE NEBULA M16

EAGLE NEBULA
MESSIER 16

The star cluster M16 hovers just on the edge of naked-eye visibility at magnitude 6.4, but is easily identified with binoculars or a small telescope. The cluster is very young, at just 5 million years old, and is still embedded in the spectacular Eagle Nebula, a site where star formation is still continuing. While Swiss astronomer Philippe Loys de Chéseaux discovered the cluster in 1745, it was Charles Messier who discovered the surrounding nebula some 19 years later. Despite its impressive appearance in photographs, the Eagle will only reveal its secrets in detail to the largest amateur telescopes.

R.A. 18h 19m, Dec. -13°47"
Magnitude 6.4
Distance 7,000 light years

PILLARS OF
CREATION

In 1995, astronomers turned the Hubble Space Telescope toward the Eagle Nebula and captured this iconic image of dense star-forming columns in the heart of the Eagle Nebula. It provided astronomers with their first detailed view of the processes involved in star formation, and was soon dubbed the "Pillars of Creation"—a name enthusiastically adopted by the media. The individual pillars in the image are up to four light years long—they mark regions whose density is high enough to resist the pressure of radiation from the bright stars in M16 forcing back the walls of gas within the nebula. Inside each of them, dozens of individual knots of gas are growing denser still, developing their own gravity so that they can pull more material from their surroundings and snowball into newborn stars.

THE SPIRE

n 2005, Hubble scientists returned
to the Eagle Nebula to study
another of its features—a longer
and more tenuous column of gas
nicknamed the Spire (of which only
the upper regions are shown here).
The spire is being rapidly eroded
by radiation from nearby newborn
stars, triggering emission of light
that surrounds it with an eerie glow.
Dense knots of material inside the
hydrogen-rich column of gas may
still develop to form stars, but as
the raw materials are stripped away
from around them their growth will
be stunted. The first generations
of stars to form in any nebula
have a natural advantage when it
comes to gaining material, and also
pull the ladder up behind them,
depriving later stellar generations
of material.

OPHIUCHUS
THE SERPENT BEARER

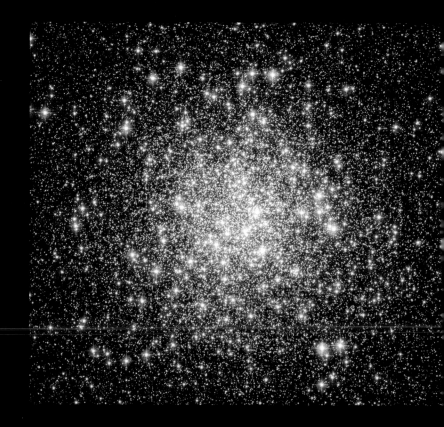

This large equatorial constellation appears rather empty of bright stars, and is best found by looking between the bright stars Altair (in Aquila) and Spica (in Virgo), and to the north of brilliant Antares in Scorpius.

Ophiuchus is intimately associated with the two-part constellation of Serpens, whose head and tail lies to either side of it. Although the constellation is usually depicted as a giant figure wrestling a snake, the figure has been seen since Roman times as Asclepius, the god of healing, who traditionally carried a staff with a serpent twined around it. The constellation is sometimes known as the "thirteenth sign of the zodiac"—the effects of precession (see page 15) mean that the ecliptic now passes through it and the Moon and planets can often be seen here.

The constellation's brightest star, Rasalhague, is a slightly variable white star of magnitude 2.1. Lying around 47 light years from Earth, it forms a binary system with an unseen companion. Rho Ophiuchi, meanwhile, is a beautiful quadruple star system wth a combined magnitude of 4.6, still embedded in the nebulosity from which it formed.

MESSIER 9 Ophiuchus contains a wealth of globular clusters, of which Messier 9 is probably the most impressive. Discovered in 1764, it lies close to the center of our galaxy, 25,800 light years from Earth, and shines at magnitude 8.4. As a result, it is only just visible through binoculars, but makes a fine sight through any telescope.

	R.A.	DEC.	TYPE	MAGNITUDE	DISTANCE
ρ Oph	16h 26m	-23°27"	Multiple star	5.0/5.9/6.7/7.3	395 l.y.
70 Oph	18h 05m	+02°30"	Binary star	4.0/6.0	17 l.y.
Barnard's Star	17h 58m	+04°42"	Red dwarf	9.5	5.9 l.y.
M10	16h 57m	-04°06"	Globular cluster	6.4	14,000 l.y.
M12	16h 47m	-01°57"	Globular cluster	7.7	16,000 l.y.

HERCULES

RASALHAGUE

α

72
71

PATH OF
BARNARD'S
STAR

κ

OPHIUCHUS

74

66
67

β

σ

70

γ

68

41

30

LOBSTER GALAXY
NGC 6240

Photographed by the Hubble Space Telescope in 2008, this unusual object is the disrupted result of a galactic collision and merger that began around 30 million years ago and may not be complete for another 100 million years. X-ray images have revealed the presence of two distinct nuclei—giant black holes some 3,000 light years apart that are ultimately destined for their own collision and merger. The galaxy pumps out far more infrared (heat) radiation than would normally be expected, perhaps because of the heating effect of accelerated star formation or due to the influence of its active black holes.

R.A. 16h 52m, Dec. +02°24"
Magnitude 12.8
Distance 400 million light years

BARNARD'S STAR
V2500 OPHIUCHI

Barnard's Star is an apparently insignificant red dwarf raised to fame by its proximity to Earth (at just 5.9 light years away) and its speed across the sky. It is the star with the highest "proper motion"— the star's own movement through space, combined with our Solar System's drift in the opposite direction, ensures that it moves half a degree (a full Moon width) across the sky every 180 years. The finder chart above shows its location within Ophiuchus and its general direction of motion. Despite its proximity, Barnard's Star was only discovered by the U.S. astronomer E.E. Barnard in 1916.

R.A. 17h 58m, Dec. +04°42"
Magnitude 9.5
Distance 5.9 light years

RHO (ρ) OPHIUCHI

Although unimpressive to the naked eye, a small telescope or even binoculars will transform Rho Ophiuchi into a beautiful multiple system with a central star of magnitude 5.0, and companions of magnitudes 5.9, 6.7 and 7.3. All the components are hot blue-white stars, just a few million years old and still embedded in remnants of the gas from which they formed. Indeed, long-exposure photographs show that the entire region is filled with beautiful glowing gas clouds and silhouetted dust lanes. in this image, Rho Ophiuchi is surrounded by blue nebulosity at the top of the picture, while the orange glow from nearby Antares intrudes at lower left.

R.A. 16h 26m Dec. -23°27"
Magnitude 4.6
Distance 395 light years

AQUILA AND SCUTUM

THE EAGLE
AND THE SHIELD

These two equatorial constellations are easily located thanks to the bright star Altair. For northern observers, Altair forms the base of the "summer triangle," but southern stargazers can find it by looking to the north of Sagittarius.

Aquila has been seen as an eagle since at least the fourth century BC—it was often associated with the bird that held Zeus's thunderbolts, or with Zeus himself, who transformed into an eagle to abduct Ganymede. Scutum, meanwhile, was invented by Johannes Hevelius in 1684, and represents the shield of Polish King John III Sobieski.

Altair, or Alpha Aquilae, is the twelfth-brightest star in the sky at magnitude 0.8. Lying just 17 light years from Earth, it is one of the few stars to have had its surface imaged directly. This revealed a bulging equator caused by the young white star's rapid spin. Magnitude-3.8 Alpha Scuti, meanwhile, is an orange giant 174 light years away, seen against some of the brightest Milky Way star clouds. Scutum also contains Messier 11, also known as the "Wild Duck," one of the Milky Way's finest star clusters.

NGC 6751 This attractive but challenging planetary nebula lies close to Lambda Aquilae in the constellation's southwest corner. Shining at magnitude 11.9, it lies some 6,500 light years from Earth and has an estimated diameter of about 0.8 light years. The nebula's complex structure is a result of collisions between fast-expanding gas from the hot central star and cooler material expelled thousands of years ago.

	R.A.	DEC.	TYPE	MAGNITUDE	DISTANCE
57 Aql	19h 55m	-8°14″	Binary star	5.3	335 l.y.
δ Sct	18h 43m	-9°03″	Variable star	c.4.7 (var)	187 l.y.
R Sct	18h 47m	-5°18″	Variable star	4.2–8.6 (var)	2,500 l.y.
M11	18h 51m	-6°16″	Open cluster	6.3	6,000 l.y.

DELPHINUS AND EQUULEUS
THE DOLPHIN AND THE LESSER HORSE

Two compact constellations—the diamond-like form of a leaping dolphin and the small skewed rectangle of a horse's head—lie to the southwest of the easily identifiable great Square of Pegasus.

Despite its small size and relative simplicity, the stars of Delphinus do indeed bear a distinct resemblance to a dolphin. Mythologically, this creature was said to be a servant of the god Poseidon, sent to rescue the bard Arion from a shipwreck. The constellation's brightest stars, Beta at magnitude 3.6 and Alpha at magnitude 3.8, are both binaries, though neither can be separated with an amateur telescope. Gamma Delphini, however, is an easily observable binary consisting of a yellow-orange primary at magnitude 4.3, and a fainter, magnitude-5.1 companion that can appear white, blue or greenish in color.

Equuleus, meanwhile, bears little resemblance to a horse of any kind, and is usually represented as simply an equine head. Despite its obscurity, the constellation dates back to ancient times and is often associated with Celeris, the foal of Pegasus.

DISTANT GIANT Globular cluster NGC 6934 in southern Delphinus was discovered by German-born British astronomer William Herschel in 1785. With a magnitude of 8.8, and a distance of 50,000 light years, it lies on the limit of binocular visiblity but is a good target for small telescopes.

	R.A.	DEC.	TYPE	MAGNITUDE	DISTANCE
Sualocin (α Del)	20h 40m	+15°55"	Binary star	3.8	240 l.y.
Rotanev (β Del)	20h 38m	+14°36"	Binary star	3.6	98 l.y.
γ Del	20h 47m	+16°07"	Binary star	4.3/5.1	104 l.y.
Kitalpha (α Equ)	21h 16m	+05°15"	Binary star	3.9	186 l.y.
ε Equ	20h 59m	+04°18"	Multiple star	5.2	197 l.y.

PE·ASUS
THE FLYING HORSE

While it bears little resemblance to the figure of a winged horse that it is supposed to represent, this constellation is still unmistakable thanks to the prominent square formed by its brightest stars.

The stars of the "Square of Pegasus" are usually depicted as the forequarters of the horse, with fainter chains of stars indicating its raised forelimbs and head. The entire depiction is upside-down for northern hemisphere observers. Of the four stars in the square, the northeastern one (Delta Pegasi) is shared with neighboring Andromeda (see page 70). Alpha, or Markab, is a white star of magnitude 2.5, 140 light years from Earth, while Beta (Scheat) is a red giant 200 light years away, varying between magnitudes 2.3 and 2.7 in an unpredictable cycle. Gamma, or Algenib, in contrast, is a hot blue star about 330 light years away, at magnitude 2.8.

The constellation's brightest star, however, is Epsilon Pegasi or Enif. This orange supergiant varies around around magnitude 2.4, and marks the horse's nose. It is 690 light years from Earth, with a blue companion easily seen through a small telescope

STEPHAN'S QUINTET When French astronomer Edouard Stephan discovered this cluster of galaxies in 1877 (now cataloged as Hickson Compact Group 92), he concluded that it contained five members. But more recent measurements have confirmed that the spiral at upper left is a foreground galaxy—just 40 million light years away compared to the 300 million light years of the four true members.

	R.A.	DEC.	TYPE	MAGNITUDE	DISTANCE
Scheat (β)	23h 04m	+28°05"	Variable star	2.3–2.7	200 l.y.
Algenib (γ)	00h 13m	+15°11"	Variable star	c.2.8	330 l.y.
Enif (ε)	21h 44m	+09°53"	Variable star	0.7–3.5	690 l.y.
51 Peg	22h 57m	+20°46"	Planetary system	5.5	50 l.y.
M15	21h 30m	+12°10"	Globular cluster	6.2	34,000 l.y.

AQUARIUS
THE WATER CARRIER

This somewhat shapeless constellation is nevertheless one of the oldest, recognized as the figure of a man pouring water from a jug by astronomers since at least the time of ancient Babylon, around 2,000 BC.

Since classical times, Aquarius has been associated with Ganymede, a beautiful youth kidnapped by Zeus (in the form of the eagle Aquila) and taken to Mount Olympus to act as cup bearer to the Greek gods. A Y-shaped group of relatively bright stars around Zeta Aquarii is known as the Water Jug, and the water pours southeast from here toward the bright star Fomalhaut in neighboring Piscis Austrinus.

The constellation's brightest stars at magnitude 2.9, Alpha and Beta Aquarii are known as Sadalmelik and Sadalsuud respectively. Both are yellow supergiants 760 and 610 light years from Earth. Beta is less luminous than Alpha, but appears fractionally brighter in Earth's skies thanks to its relative proximity. Zeta Aquarii, meanwhile, is an attractive double star with near-twin white components of magnitudes 4.3 and 4.5, sitting

SATURN NEBULA This planetary nebula, catalogue`d as NGC 7009, lies around 5,200 light years from Earth in western Aquarius, and takes its name from its obvious resemblance, seen through a small telescope, to the ringed planet Saturn. The compact nature of this magnitude-8.0 nebula makes it relatively easy to observe.

	R.A.	DEC.	TYPE	MAGNITUDE	DISTANCE
ζ Aquarii	22h 29m	-00°01″	Binary star	4.3/4.5	103 l.y.
M2	21h 34m	-00°49″	Globular cluster	6.5	37,500 l.y.
NGC 7009	21h 04m	-11°22″	Planetary nebula	8.0	5,200 l.y.
NGC 7293	22h 30m	-20°48″	Planetary nebula	7.3	450 l.y.

AQUARIUS THE HELIX NEBULA NGC 7293

CETUS

THE SEA MONSTER

This large constellation, straddling the celestial equator southwest of Taurus, can be challenging to spot on account of its tendency to change appearance depending on the brightness of one of its key stars.

Strictly speaking, the Latin word *cetus* translates as "whale," but since ancient times this constellation has been identified as a far more fearsome sea monster called Typhon. This creature plays a role in myths associated with the constellations Pisces and Capricornus, but is best known as the monster sent by Hera, the jealous Queen of the Gods, to ravage the kingdom of Cassiopeia and Cepheus.

Cetus is home to a variety of interesting objects including the attractive binary star Gamma Ceti, the line-of-sight double Alpha Ceti (Menkar) and the nearby stars Tau Ceti and UV Ceti A and B. However, its most famous star by far is Omicron Ceti, or Mira—a slowly pulsating red giant that varies between around magnitude 3.0 and magnitude 10.0 in a 332-day cycle. At its brightest, Mira joins the head and body of Cetus at the neck, but at its faintest, its disappearance splits Cetus into two distinct groups.

SPEEDING WONDER This ultraviolet image reveals that Mira (whose name means "wonderful") is racing through space at high speed and leaving a trail of hot gas in its wake. This material has been flung off in each of its long, slow pulsations, and pulled away by the tug of a close binary companion. Theoretical models suggest the star varies between 300 and 400 times the diameter of the Sun in the course of each 332-day cycle.

	R.A.	DEC.	TYPE	MAGNITUDE	DISTANCE
γ **Ceti**	02h 43m	+03°14"	Binary star	3.5	82 l.y.
Mira (ο)	02h 19m	-02°59"	Variable star	3.0–10.0 (var)	420 l.y.
τ **Ceti**	01h 55m	-15°56"	Main sequence	3.5	12 l.y.
UV Ceti	01h 39m	-17°57"	Dwarf binary star	8.6	12.5 l.y.
M77	02h 43m,	00°00"	Active galaxy	9.6	60 million l.y.

ORION
THE HUNTER

This bright and unmistakable constellation dominates both northern and southern hemisphere skies early in the year. It is studded with celestial marvels ranging from turbulent starbirth nebulae to dying stars near the end of their lives.

Orion represents a mighty hunter of Greek and Roman myth, the lover of Artemis, goddess of the hunt. Orion was said to have boasted that he could overcome any creature on Earth, whereupon the Earth brought forward a scorpion that stung him to death. For this reason, Orion and Scorpius are today preserved on opposite sides of the sky.

Two of the brightest stars in the sky, Betelgeuse and Rigel, mark the hunter's shoulder and knee, while a chain of three somewhat fainter ones mark his belt. The easternmost of these, magnitude-1.8 Zeta Orionis, or Alnitak, is a multiple star that reveals a companion of magnitude 4.0 with a small telescope, and a third component of magnitude 9.5 through larger instruments. South of Alnitak, runs a chain of stars and nebulosity forming Orion's Sword—one of the most rewarding regions in the entire sky.

HORSEHEAD NEBULA A vast cloud of interstellar material lies behind much of Orion, only becoming visible when excited by the light of nearby stars, or silhouetted against more distant light sources. Perhaps the most famous region of the cloud is the Horsehead Nebula, or Barnard 33, a renowned celestial chesspiece formed by a loop of obscuring dust 3.5 light years wide, visible against the IC 434 nebula.

	R.A.	DEC.	TYPE	MAGNITUDE	DISTANCE
Betelgeuse (α)	05h 55m	+07°24"	Variable star	0.3–1.2	640 l.y.
Mintaka (δ)	05h 32m	-00°18"	Multiple star	2.2/6.9/14.0	900 l.y.
Alnitak (ζ)	05h 41m	-01°47"	Multiple star	1.8/4.0/9.5	820 l.y.
Trapezium Cluster (θ)	05h 35m	-05°27"	Open cluster	4.0	1,500 l.y.
σ Ori	05h 39m	-02°36"	Multiple star	3.8/6.6/9.0	1,150 l.y.
M42, M43	05h 35m	-05°27"	Diffuse nebula	4.0	1,350 l.y.
M78	05h 47m	+00°03"	Diffuse nebula	8.3	1,600 l.y.
Barnard 33	05h 41m	-02°28	Dark nebula	N/A	1,500 l.y.

ORION INSIDE VIEW

BETELGEUSE
α ORIONIS

Alpha Orionis is the distinctive red supergiant Betelgeuse. Despite its "alpha" designation, and its rank as the tenth-brightest star in the sky, it is not the brightest star in Orion—that honor goes to Rigel. Betelgeuse is a fascinating star, however: it is so large that it is unstable and fluctuates in both size (between 300 and 400 times the diameter of the Sun) and brightness (between magnitudes 1.2 and 0.3). The huge size of Betelgeuse also allowed the Hubble Space Telescope to perform a notable first—direct imaging of the surface of a distant star.

R.A. 05h 55m, Dec. +07°24″
Magnitude 0.3–1.2 (var)
Distance 640 light years

LL ORIONIS

This faint young star originated in the Orion Nebula relatively recently, and is now moving through the nebula's outer reaches, drifting away into space. An impetuous youth, LL Orionis generates a much fiercer stellar wind than more sedate stars like the Sun. As this wind encounters the oncoming pressure of gas in the wider nebula, it generates a curving shock wave called a "bow shock," analogous to the wave generated in front of a fast-moving ship. Collisions between particles from the nebula and those from LL Orionis's stellar winds release the energy that makes the interface glow.

R.A. 05h 35m, Dec. -05°25″
Magnitude 11.5
Distance 1,350 light years

◄ FLAME NEBULA
NGC 2024

Lying just a little way to the north of the famous Horsehead Nebula, the Flame is another celebrated cloud of interstellar gas and dust. It is illuminated in part by the radiation that floods out from Alnitak or Zeta Orionis, the easternmost star of Orion's Belt, and in part by the radiation from new stars being born inside the nebula. Dark clouds of dust lying in front of the nebula give it the appearance of a billowing flame in visible light, but this near-infrared view from the European Southern Observatory's VISTA telescope lifts the veil of dust to reveal a complex celestial landscape beneath.

R.A. 05h 42m, Dec. -01°51"
Magnitude 2.0
Distance c.900 light years

RIGEL
β ORIONIS

Despite its designation, Rigel is the constellation's brightest star. In striking contrast to Betelgeuse, it is a brilliant blue-white color. Some have speculated that the mislabeling of the two stars suggests Betelgeuse actually outshone Rigel in the historic past. Brilliant blue-white supergiants such as Rigel are some of the most massive stars known—Rigel has an estimated mass of 17 Suns, a surface temperature of 19,800°F (11,000°C) and a luminosity 66,000 times that of our own Sun. Rigel has a close companion star in the sky, but this is merely a chance alignment between unrelated stars.

R.A. 05h 14m, Dec. -08°12"
Magnitude 0.1
Distance 860 light years

ORION THE ORION NEBULA M42

INFANT STARS

As one of the closest nearby star factories, the Orion Nebula offers a unique opportunity for astronomers to test their theories and look at the properties of newborn stars. One important feature they have identified is the near-ubiquity of protoplanetary discs or "proplyds." These dense clouds of gas and dust persist in orbit around stars even after they have begun to shine, and therefore seem to be an ideal source of raw materials for the later formation of solar systems. Another significant Hubble discovery within the nebula is an abundance of lightweight brown dwarfs—"failed stars" that never achieved sufficient mass to ignite nuclear fusion reactions, but still glow with heat from their formation.

R.A. 05h 35m, Dec. -05°27"
Magnitude 4.0
Distance 1,350 light years

TRAPEZIUM CLUSTER

At the heart of the Orion Nebula lies a tight cluster of stars that smaller telescopes show as a quadruple star system. The cluster, universally known as the Trapezium, was first reported by the Italian astronomer Galileo Galilei in his 1610 book *The Starry Messenger*, and while the number of stars in the cluster has steadily increased, the name has stuck. The five dominant stars of the Trapezium, shown in this Hubble Space Telescope image, each have masses between 15 and 30 times that of the Sun. This ensures that they will live fast and die young, but for the moment the ultraviolet radiation they pour out into the surrounding nebulosity does much to illuminate the gas clouds.

▶ GREAT NEBULA

M42 is easily seen wth the naked eye as a patch of fuzzy light just south of the multiple star system Theta Orionis in Orion's Sword. Many of the nebula's features can already be traced using just binoculars or a small telescope, although the nebula appears greenish rather than pink. Larger telescopes only increase the levels of detail visible in this wonderful flower-like structure. As well as M42, Orion's Sword is home to several other interesting nebulae, though none is as bright and large as the so-called Great Nebula.

MONOCEROS
CANIS MINOR

THE UNICORN AND THE LESSER DOG

Lying directly to the east of brilliant Orion, the large "W" of Monoceros provides a much fainter observing prospect. Fortunately, Procyon, the brightest star of Canis Minor, lies directly to its north.

Monoceros stretches across a bright region of the Milky Way, which makes its pattern all the more difficult to distinguish. It is widely thought to have been invented by Dutch theologian Petrus Plancius around 1613, although some early historians claimed it had more ancient Persian or Arabic origins. Despite its obscurity, Monoceros is home to several interesting stars and deep-sky objects—Beta Monocerotis, for instance, is an attractive triple star visible through the smallest telescopes.

Canis Minor, in contrast, is a compact region with little to offer apart from the bright star Procyon shining at magnitude 0.3. Despite this, the lesser dog is one of Ptolemy's classical constellations, and has been recognized since the first century BC. Procyon itself is one of the closest bright stars to Earth, just 11.4 light years away and similar in many ways to the even brighter Sirius (see page 142).

V838 MONOCEROTIS In January 2002, astronomers spotted an unusual stellar explosion to the southwest of Delta Monocerotis—an object soon designated as V838 Monocerotis. Since then, this strange star has been a subject of intensive study as the "light echo" of the explosion has reflected off surrounding gas and dust. The origin of the outburst, it seems, was a supergiant star nearing its death in a supernova.

	R.A.	DEC.	TYPE	MAGNITUDE	DISTANCE
β Mon	06h 29m	-07°02"	Multiple star	3.9	690 l.y.
Procyon (α CMi)	07h 39m	+05°13"	Binary star	0.3	11.4 l.y.
M50	07h 03m	-08°20"	Open cluster	5.9	3,200 l.y.
NGC 2244	06h 32m	+04°52"	Open cluster	4.8	5,200 l.y.
NGC 2264	06h 41m	+09°53"	Open cluster	3.9	2,700 l.y.

HUBBLE'S VARIABLE NEBULA, NGC 2261

This curious faint nebula lies close to 15 Monocerotis in the north of the constellation and is noted for its strange behavior, varying in both brightness and apparent shape. It is named in honor of the great U.S. astronomer Edwin Hubble, who was the first person to study it in detail, and it is associated with the unstable variable star R Monocerotis, seen at the bottom of this Hubble Space Telescope image. R Mon is a T Tauri star—an unstable stellar newborn that is stll expelling excess material and fluctuating in brightness. Since the nebula is illuminated by reflection of light from R Mon, it varies along with the star, albeit with some delay due to the time light takes to travel along its length.

R.A. 06h 39m, Dec. +08º44"
Magnitude 10.0–12.0 (var)
Distance 2,500 light years

ROSETTE NEBULA NGC 2244

This beautiful nebula, to the east of Epsilon Monocerotis, looks like nothing so much as a flower blossoming in space. Officially, NGC 2244 is the star cluster that lies in the middle of all the nebulosity. Packed with heavyweight blue stars, its radiation excites and illuminates the surrounding gas while stellar winds "hollow out" the central region. NGC 2237, 2238, 2239 and 2246 are the designations for various parts of the Rosette itself. It lies around 5,200 light years from Earth and has a diameter of around 130 light years. While the nebula itself is best spotted using a telescope with low magnification, the star cluster is visible to the naked eye and a good subject for binocular exploration.

R.A. 06h 34m, Dec. +05º00"
Magnitude 9.0
Distance 5,200 light years

▶ CONE NEBULA
NGC 2264

Two related objects share the
designation of NGC 2264—the
bright Chrismas Tree star cluster,
and a nearby pillar of opaque
dust, the Cone Nebula. A nearby
emission nebula provides an
attractive background to both. The
star cluster is easily found with the
naked eye or binoculars, and forms
a roughly conical shape of its own,
which meets the dark cone of gas
and dust tip-to-tip. The cone is in
fact a "pillar of creation"—part of a
star-forming region that once gave
rise to the Christmas Tree Cluster
itself, and has now been beaten
back to a shadow of its former self
by radiation from its offspring.

R.A. 06h 41m, Dec. +09°53"
Magnitude 3.9
Distance 2,700 light years

CANIS MAJOR
THE GREAT DOG

Lying to the southeast of Orion, this constellation is unmistakable thanks to the presence of Sirius, the brightest star in the entire sky. However, Canis Major also encompasses other bright stars and deep-sky objects as it is crossed by a rich swathe of the Milky Way.

Canis Major was probably first associated with a dog because of the way its stars faithfully pursue the great hunter Orion across the sky. In particular, although Sirius's name derives from the ancient Greek for "scorching," it has long been known as the "Dog Star." This hot white star, with roughly twice the mass of the Sun, is 25 times as luminous, but owes its brightness in our skies largely to proximity—at a distance of 8.6 light years, it is the fifth-closest star system to Earth.

In contrast, some of the constellation's other bright stars are far more distant, and truly brilliant in their own right—Delta Canis Majoris or Wezen, for example, shines at magnitude 1.8 over a distance of 1,800 light years thanks to an intrinsic luminosity of around 47,000 Suns.

THE SIRIUS SYSTEM Despite the Dog Star's proximity to Earth, its binary companion, Sirius B, is tiny and hard to spot. At magnitude 8.5, it is not terribly faint, but its light is simply drowned out by the brilliance of its neighbor. For all its tiny size and faintness, Sirius B has a mass equal to that of the Sun—it is a white dwarf, the burnt-out core of a star that once outshone Sirius A.

	R.A.	DEC.	TYPE	MAGNITUDE	DISTANCE
Sirius (α)	06h 45m	-16°43"	Binary star	-1.46	8.6 l.y.
Mirzam (β)	06h 23m	-17°57"	Variable star	2.0	500 l.y.
M41	06h 46m	-20°44"	Open cluster	4.5	2,300 l.y.
NGC 2362	07h 19m	-24°57"	Open cluster	4.1	5,000 l.y.

TAU (τ) CANIS MAJORIS CLUSTER, NGC 2362

This young and heavyweight open cluster surrounds the magnitude-4.4 blue supergiant Tau Canis Majoris. The cluster contains the mass of more than 500 Suns and has a combined apparent magnitude of 4.1, making it an easy naked-eye object and a rewarding sight through binoculars or a small telescope. The Spitzer Space Telescope infrared image above was used to search for evidence of protoplanetary disks around these young stars, and revealed that they only persist around the smallest stars in the cluster.

SUPERGIANT STAR VY CANIS MAJORIS

This red supergiant star just to the south of the Tau Canis Majoris cluster is among the largest stars known. Dropped into the middle of our Solar System, its outer layers would stretch beyond the orbit of Saturn. With 40 times the mass of the Sun, VY is hurtling toward its eventual doom in a supernova explosion. During these final stages of its life it has become an unstable variable, flinging away vast amounts of dust and gas from its outer layers in unpredictable outbursts. This colorful view from the Hubble Space Telescope maps the distribution of dust around the star using polarized light.

THOR'S HELMET NGC 2359

This beautiful bubble in space is an emission nebula, created by high-energy stellar winds from the blue supergiant star at the bubble's center. As these particles plow into surrounding material in interstellar space, they create a glowing, roughly spherical shock wave around 30 light years across. As the star moves through space, it carries the bubble with it, and movement through the surrounding nebulosity creates a curving "bow shock" that sweeps material back around the sides of the nebula and gives it a strong resemblance to a winged helmet.

▶ DANCING SPIRALS NGC 2207/IC 2163

A little way to the south of Beta Canis Majoris lies this attractive pair of galaxies, discovered by English astronomer John Herschel in 1835. The two spirals have been engaged in a slow process of collision for some 40 million years, flinging long streamers of stars out into intergalactic space and triggering bursts of new star formation as their tidal forces tug at each other. Eventually they are doomed to merge, probably forming an elliptical galaxy in a process that will take another 100 million years or more.

R.A. 07h 19m, Dec. -24°57″
Magnitude 4.1
Distance 5,000 light years

R.A. 07h 22m, Dec. -25°46″
Magnitude 6.5–9.6 (var)
Distance 3,800 light years

R.A. 07h 19m, Dec. -13°13″
Magnitude 11.5
Distance 15,000 light years

R.A. 06h 16m, Dec. -21°22″
Magnitude 12.2, 11.6
Distance 114 million light

HYDRA
THE WATER SNAKE

The largest constellation in the sky, Hydra's faint stars wind their way around a huge swathe of the celestial equator to the south of more prominent constellations, such as Leo and Virgo.

This huge and ancient constellation was originally depicted as a serpent by astronomers in ancient Babylon. Although its name recalls the many-headed Lernaean Hydra fought by Hercules, it is more often seen as one player in a story linking the nearby constellations of Corvus and Crater (see page 150). Hydra's extensive size and faintness make it hard to identify—only its head and neck, to the southwest of Leo, are reasonably prominent.

The constellation's brightest star, Alphard, is a red giant some 180 light years from Earth. It has a name that means "solitary one" and, at magnitude 2.0, is indeed the brightest object in a relatively barren region of the sky. Deep-sky objects include the open cluster Messier 48, just below the serpent's head (a good target for binoculars or a small telescope), and the spectacular Southern Pinwheel Galaxy Messier 83 in the constellation's tail.

GALACTIC SILHOUETTE NGC 3314, near Hydra's southern boundary, is a rare and beautiful cosmic coincidence—a pair of perfectly aligned galaxies 117 million and 140 million light years from Earth. Light from the more distant galaxy turns the foreground spiral largely transparent, giving us a unique look at its dusty skeleton.

	R.A.	DEC.	TYPE	MAGNITUDE	DISTANCE
ε **Hya**	08h 47m	+06°25″	Multiple star	3.4/7.5	130 l.y.
P **Hya**	13h 30m	-23°17″	Variable star	3.2–11.0	2,000 l.y.
M48	08h 14m	-05°48″	Open cluster	5.5	1,500 l.y.
M83	13h 37m	-29°52″	Spiral galaxy	7.6	15 million l.y.
NGC 3242	10h 25m	-18°38″	Planetary nebula	7.7	2,500 l.y.

BUCKLED SPIRAL
ESO 510-G13

This strange galaxy to the west of Pi Hydrae may only be visible through the largest amateur telescopes, but is fascinating nevertheless. It is clearly a spiral galaxy lying more or less edge-on to Earth, with a dark outer dust lane silhouetted against the background glow of its starry disk and hub, but for some reason the disk had become warped or bent out of shape. Astronomers believe the distortion is caused by gravitational forces at work as ESO 510-G13 swallows up and cannibalizes a smaller galaxy. Intriguingly, there are recent suggestions that our own Milky Way galaxy might have a similar warp on its outer edge.

MESSIER 68

Hydra's only globular cluster lies midway along the line between Beta and Gamma Hydrae. Lying on the opposite side of the sky from the crowded galactic center, it is among the more isolated globulars in orbit around our galaxy, with an estimated distance of 33,000 light years from Earth. Too faint to see with binoculars, Messier 68 was discovered by Charles Messier himself in 1780, and is an object better suited to larger telescopes. These reveal a very loose, open structure with individual stars resolvable almost all the way to the center of the cluster.

SOUTHERN PINWHEEL
MESSIER 83

This beautiful barred spiral is undoubtedly Hydra's most alluring deep-sky object. Roughly one-third the size of the full Moon but quite bright at magnitude 7.6, it can be located with binoculars on the constellation's southern edge, to the south of R Hydrae in the serpent's tail, and is a rewarding sight through larger instruments. M83 is a barred spiral about half the size of the Milky Way, with prominent dust lanes that can be traced all the way to the core. It was discovered in 1752 by Nicolas Louis de Lacaille, the French astronomer responsible for naming many of the southern constellations.

▶ ARC OF STARBIRTH

This stunning image from the Hubble Space Telescope highlights the glowing fires of star formation lighting up one of Messier 83's spiral arms. The galaxy is forming stars at a much faster rate than our own Milky Way, perhaps due to the gravitational influence of nearby galactic neighbors. The image captures key moments in the life cycle of stars, ranging from their birth in dark clouds and bright starbirth nebulae, through their emergence as brilliant open clusters, to their eventual deaths, in the form of some 60 supernova remnants that NASA scientists have counted within the picture.

R.A. 13h 55m, Dec. -26°46¨
Magnitude 13.4
Distance 150 million light years

R.A. 12h 39m, Dec. -28°45¨
Magnitude 9.7
Distance 33,000 light years

R.A. 13h 37m, Dec. -29°52¨
Magnitude 7.6
Distance 15 million light years

CORVUS, CRATER, AND SEXTANS

THE CROW, THE CUP, AND THE SEXTANT

Three small constellations, two of which are ancient and relatively easy to spot, run along the back of the great water snake Hydra, separating it from the zodiac constellations of Virgo and Leo.

In mythology, Corvus and Crater are associated with Hydra in a curious story of the god Apollo. According to this tale, the great god sent his servant the crow to fill his cup with water from a well, but the bird was distracted by a ripe fig tree. On his return, he clutched a water snake that he claimed had been blocking the well, but Apollo saw through the lie, and threw the cup, crow, and serpent into the sky in anger. Sextans, in contrast, is a relatively modern and obscure constellation invented by the Polish astronomer Johannes Hevelius in the late 17th century.

Most of the stars in these three constellations are unremarkable, but Delta Corvi (also known as Algorab) is worthy of attention—a small telescope reveals that the magnitude-3.0 primary star has a companion of magnitude 8.5, which often displays an unusual purple color.

ANTENNAE GALAXIES This strange pair of galaxies lies close to the border of Corvus with Crater, an estimated 45 million light years from Earth. Wide-angle views show long trails of stars resembling an insect's antennae, but detailed images reveal that Antennae are two spiral galaxies (NGC 4038 and 4039) in the process of collision.

	R.A.	DEC.	TYPE	MAGNITUDE	DISTANCE
Algorab (δ Crv)	12h 30m	-16°31"	Binary star	3.0/8.5	88 l.y.
NGC 3115	10h 05m	-07°43"	Lenticular galaxy	9.9	32 milion l.y.
NGC 4038/4039	12h 02m	-18°52"	Colliding galaxies	10.3	45 million l.y.

CENTAURUS
THE CENTAUR

This sprawling constellation is one of two in the sky that represent mythical centaurs. Covering a large area of sky between Hydra and the Southern Cross, it contains several of the highlights of southern skies.

Although it was seen as a bull by ancient Babylonian astronomers, by classical times this group of stars was firmly established as a centaur, and usually associated specifically with Chiron, wise tutor of the warrior Achilles. The centaur's body is formed by a rough pentagon of stars, with the brightest stars of all lying to the southeast.

Alpha Centauri, or Rigil Kentaurus, is the third-brightest star in the sky at magnitude -0.1. Small telescopes reveal that it is a binary consisting of two stars, each very similar to the Sun. An outlying third member of the system, the magnitude-11.0 red dwarf Proxima Centauri, is the closest star to the Sun at a distance of 4.26 light years. Beta Centauri (Hadar) is also a triple star system, but in this case one that consists of heavyweight and highly luminous blue stars some 525 light years from Earth (only two of which can be visually separated).

STARS OF THE CENTAUR This stunning view of the southern Milky Way shows the extensive constellation of Centaurus. Alpha and Beta Centauri form a spectacular pair at lower left, alongside the Southern Cross. In 2012, astronomers announced the discovery of an Earth-sized planet orbiting the fainter of Alpha Centauri's near-twin stars.

	R.A.	DEC.	TYPE	MAGNITUDE	DISTANCE
Rigil Kentaurus (α)	14h 40m	-60°50"	Multiple star	-0.1	4.4 l.y.
Hadar (β)	14h 04m	-60°22"	Multiple star	0.6	525 l.y.
NGC 5139	13h 27m	-47°29"	Globular cluster	3.7	16,000 l.y.
NGC 5128	13h 26m	-43°01"	Active galaxy	7.0	15 million l.y.
NGC 3918	11h 50m	-57°11"	Planetary nebula	8.1	2,600 l.y.

RADIO JETS

The bright galaxy NGC 5128 is easily spotted with binoculars and appears to be an elliptical ball of stars through small telescopes. Slightly larger instruments or long-exposure photographs will reveal that this bright disk is bisected by a dark stream of dust. NGC 5128 coincides with a strong radio source known as Centaurus A, and as a result the galaxy itself is informally known by this name as well. Centaurus A is one of the closest and brightest active galaxies to Earth—multiwavelength composite images such as the one above reveal twin jets of material emerging in opposite directions from the bright nucleus, emitting both radio waves (shown in orange) and X-rays (in blue).

R.A. 13h 26m, Dec. -43°01″
Magnitude 7.0
Distance 15 million light years

▶ STARBURST REGION

Elliptical galaxies normally lack the raw material to fuel ongoing star formation, and so tend to be dominated by long-lived, lower-mass stars. The injection of new material into Centaurus A, however, has triggered an enormous new wave of starbirth around and within the dark band. This Hubble Space Telescope image reveals the telltale pinkish glow of huge starbirth nebulae, and the brilliant blue of young star clusters emerging from their cocoons. The multiwavelength view combines visible light with near-infrared and ultraviolet images to render much of the material in the galaxy's dust lane transparent.

CANNIBAL GALAXY

This unusual image combines a
near-infrared view of Centaurus A
(background) with a map derived
from radio observations of carbon
monoxide molecules within the
galaxy. It reveals an almost perfect
trapezium of gas-rich material
embedded within the galaxy and
coinciding with the visible dust lane.
Different colors on opposite sides of
the nucleus show regions that are
moving toward Earth (green)
or away (red)—evidence that the
entire structure is rotating. This
strange feature is the surviving
skeleton of a gas- and dust-rich
galaxy (probably a spiral similar
to our own Milky Way) that has
been engulfed by a largely gasless
elliptical. The blaze of light from
the center of the infrared image
indicates the precise position
of the galaxy's active nucleus,
a supermassive black hole that
has been stirred into life by these
tumultuous events.

OMEGA CENTAURI

Bright enough to be given its own Greek "star letter," Omega Centauri is the largest and brightest globular cluster in the heavens. Easily visible to the naked eye as a patch of light about the size of the full Moon, it shines at an impressive magnitude 3.7. In part, this is due to its proximity—at 16,000 light years it is one of the closer globulars to Earth. But Omega is also a genuinely impressive object in its own right, an enormous stellar conglomerate of several million stars crowded into a region just 170 light years across. Any optical instrument will begin to show individual stars in the cluster's outer edges, but only the largest instruments can resolve the blizzard of stars near its center.

R.A. 13h 27m, Dec. -47º29"
Magnitude 3.7
Distance 16,000 light years

A SKY FULL OF STARS

This Hubble Space Telescope view peers into the crowded heart of Omega Centauri, capturing tens of thousands of stars in a single image. Stars within globular clusters are extremely old—they formed in a much earlier phase of cosmic evolution and are only so long-lived because they lack metal "pollutants"—heavier elements that accelerate the rate at which more recently formed stars burn their fuel.

Evidence suggests Omega's stars formed in several stages over some 2 billion years, concluding 10 billion years ago. In contrast, most globulars seem to have formed in a single burst. This lends weight to a theory that Omega is not a normal globular at all, but is actually the dense core of a dwarf galaxy cannibalized by the Milky Way.

▶ **STELLAR VARIETY**

This detailed view from Hubble's Wide Field Camera 3 shows a colorful array of stars in the heart of Omega Centauri. Yellow stars, representing the vast majority of the cluster's population, are still in the midst of their prolonged, sedate middle age. Red stars are giants that have swollen and brightened as they near the end of their lives, while faint blue-white dots are white dwarfs—the burnt-out cores of stars that have long since died. The bright blue stars are known as "blue stragglers." In theory, there is no place for short-lived heavyweight stars in such a cluster—astronomers think they probably formed relatively recently through the collision and merger of "normal" stars in the cluster's crowded heart.

LUPUS
THE WOLF

This bright but somewhat shapeless constellation is best located by looking between Antares in Scorpius and the bright pair of Alpha and Beta Centauri. Lying across a bright region of the Milky Way, it contains a variety of interesting objects.

Although this star grouping dates back at least to classical Greece, it was not identified as a wolf until medieval times. Earlier astronomers knew it as Bestia or Fera, both words that simply mean "wild animal." The beast was often shown impaled on a spear wielded by neighboring Centaurus.

The constellation's brightest star, Alpha Lupi or Kakkab, is a blue stellar giant shining at magnitude 2.3 across 550 light years. Mu Lupi, meanwhile, is an attractive multiple star—small telescopes reveal a magnitude-7.2 companion in orbit around a magnitude-4.3 primary, while larger instruments split the primary into twin components of magnitudes 5.1 and 5.2. Numerous star clusters are scattered through the constellation, with NGC 5822 the brightest of the open clusters, and NGC 5986, at magnitude 7.1, the brightest of several globulars.

RETINA NEBULA Catalogued as IC 4406, this unusual, apparently rectangular cloud of gas is in fact a planetary nebula—an expanding shell of gas puffed out into space from the outer layers of a dying Sunlike star. The Retina's unique shape is due to the particular angle at which we see it—in reality, the nebula is roughly doughnut-shaped, but we are viewing it from almost exactly edge-on.

	R.A.	DEC.	TYPE	MAGNITUDE	DISTANCE
γ Lupi	15h 35m	-41°10"	Binary star	2.5	570 l.y.
μ Lupi	15h 19m	-47°53"	Multiple star	5.1/5.2/7.2	290 l.y.
NGC 5822	15h 04m	-54°24"	Open cluster	6.5	6,000 l.y.
NGC 5986	15h 46m	-37°47"	Globular cluster	7.1	34,000 l.y.
IC 4406	14h 22m	-44°09"	Planetary nebula	10.5	2,000 l.y.

SCORPIUS
THE SCORPION

This large and bright zodiac constellation, crossing the southern Milky Way between Sagittarius and Libra, is easily identified thanks to the presence of blood-red Antares, the so-called "rival of Mars."

Scorpius is one of the oldest constellations, found in even the earliest astronomical records. While its twisting form does indeed bear some resemblance to a scorpion, its modern pattern is truncated at the head, with the claws forming nearby Libra.

Shining at an average of magnitude 1.0, Antares is subject to slow variations in its light and is one of the most extreme stars known—dropped into the middle of the Solar System, it would engulf all the planets out to Jupiter. Antares has a close companion of magnitude 5.5 visible in medium-sized telecopes, but the constellation has some easier multiple targets. Acrab (Beta Scorpii) reveals blue components of magnitudes 2.6 and 4.9 through small telescopes, while Jabbah (Nu Scorpii), is a "double double" that reveals two components with a small telescope, and four with a medium-sized instrument.

BUTTERFLY CLUSTER This contrasting star cluster in eastern Scorpius is visible to the naked eye as a concentrated patch of light in the Milky Way, roughly the size of the full Moon. Cataloged as Messier 6, it lies around 1,600 light years from Earth. Though dominated by hot blue stars, its brightest star is the orange giant BM Scorpii.

	R.A.	DEC.	TYPE	MAGNITUDE	DISTANCE
Antares (α)	16h 29m	-26°26"	Variable star	0.9–1.2	600 l.y.
Acrab (β)	16h 05m	-19°48"	Multiple star	2.6/4.9	530 l.y.
Jabbah (ν)	16h 12m	-19°28"	Multiple star	3.9	440 l.y.
ξ Sco	16h 04m	-11°22"	Multiple star	4.2	93 l.y.
M4	16h 24m	-26°32"	Globular cluster	5.9	7,200 l.y.
M7	17h 54m	-34°49"	Open cluster	3.3	800 l.y.

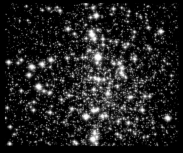

MESSIER 80

This globular cluster is among the densest known, containing hundreds of thousands of stars in a ball with a diameter of just 95 light years. It is easily located with binoculars or a small telescope, midway between Antares and Beta Scorpii, but even the largest instruments will be unable to resolve detail in its crowded core. Perhaps unsurprisingly for such a dense cluster, M80 contains a large number of "blue stragglers," unexpectedly bright and hot stars. These are thought to form through collisions and mergers among the more sedate stars of a cluster's general population.

R.A. 16h 17m, Dec. -22º59˝
Magnitude 7.9
Distance 32,600 light years

MESSIER 4

In contrast to M80, the globular cluster Messier 4 shows a relatively loose structure, as displayed in this Hubble Space Telescope view of its central regions. The cluster is on the edge of naked-eye visibility, but is easily spotted with binoculars or a small telescope, lying directly to the west of Antares. Slightly larger instruments will begin to resolve individual stars across the cluster's 75-light-year diameter. At a distance of just 7,200 light years, M4 is one of the closest globular clusters to Earth. Studies of the cluster have identified white dwarfs within it whose lives began around 13 billion years ago—the oldest stars associated with the Milky Way.

R.A. 16h 24m, Dec. -26º32˝
Magnitude 5.9
Distance 7,200 light years

PTOLEMY CLUSTER
MESSIER 7

This knot of bright stars in the Milky Way close to the scorpion's sting is named after the Greek-Egyptian astronomer Ptolemy of Alexandria, who wrote the first description of it around AD 130. Naked-eye stargazers may have difficulty picking it out against the background star clouds of this region, but binoculars or a low-powered telescope will show the cluster in all its glory. Unlike M4 and M80, Messier 7 is an open cluster. Its loose collection of several dozen stars formed around 200 million years ago and is dominated by bright blue stars and a single red giant.

R.A. 17h 54m, Dec. -34º49˝
Magnitude 3.3
Distance 800 light years

▶ ANTARES
ALPHA (α) SCORPII

Brilliant red Antares is normally the sixteenth-brightest star in the sky, a highly evolved red supergiant that varies slowly and erratically between magnitudes 0.9 and 1.2. It is the brightest member of the Scorpius–Centaurus OB Association, a nearby group of massive stars, and is also the group's heaviest member. With a mass of 17 Suns, Antares has squandered its supplies of nuclear fuel at a tremendous rate, and is now hurtling toward destruction, despite being a mere 12 million years old. As this stunning image shows, Antares lies within a dense region of nebulosity, revealed in long-exposure photographs.

R.A. 16h 29m, Dec. -26º26˝
Magnitude 0.9–1.2 (var)
Distance 600 light years

SAGITTARIUS
THE ARCHER

Lying in one of the richest regions of the Milky Way, this constellation is strewn with numerous deep-sky objects ranging from nebulae to star clusters and the very center of our galaxy itself.

The brightest stars of Sagittarius form a distinctive "teapot" shape in front of dense Milky Way starfields. Several ancient civilizations saw it as a horse and rider, but the ancient Greeks saw it as a centaur wielding a bow and arrow. The brightest star is Epsilon Sagittarii (Kaus Australis), a "white giant" star of magnitude 1.8 some 145 light years from Earth. Beta Sagittarii or Arkab, meanwhile, is a line-of-sight double consisting of blue and white stars at magnitudes 4.0 and 4.3, 380 and 140 light years away respectively. A small telescope will show a third component of magnitude 7.1.

Deep-sky highlights within Sagittarius include the Lagoon, Omega, and Trifid nebulae (M8, M17 and M20). The heart of the Milky Way, marked by the radio source Sagittarius A*, lies 26,000 light years away close to Gamma Sagittarii, but is hidden behind dense intervening star clouds.

TRIFID NEBULA Dark canyons of dust divide this glowing gas cloud into three roughly equal parts and give the Trifid Nebula, Messier 20, its name. It lies around 5,200 light years from Earth and presents a rare and beautiful combination of a pinkish emission nebula, a blue reflection nebula and a star cluster that provides the radiation to illuminate them both, glowing at just below naked-eye visibility.

	R.A.	DEC.	TYPE	MAGNITUDE	DISTANCE
Kaus Australis (ε)	18h 24m	-34°23"	Binary star	1.8	145 l.y.
Arkab (β)	19h 23m	-44°27"	Visual double	4.0, 4.3	380, 140 l.y.
RY Sgr	19h 17m	-33°31"	Variable star	6.0–7.1 (var)	9,000 l.y.
M8	18h 04m	-24°23"	Diffuse nebula	6.0	5,000 l.y.
M17	18h 20m	-16°11"	Diffuse nebula	6.0	5,500 l.y.
M20	18h 02m	-23°02"	Diffuse nebula	6.3	5,200 l.y.
M22	18h 36m	-23°54"	Globular cluster	5.1	10,400 l.y.
M23	17h 57m	-19°01"	Open cluster	6.9	2,100 l.y.

SAGITTARIUS INSIDE VIEW

ALIEN VISITOR
MESSIER 54

At first glance this globular cluster seems to be a typical example of its type, but Messier 54 hides an extraordinary secret. Although the cluster has been recognized since Charles Messier first cataloged it in 1778, it was only in 1994 that astronomers analyzing the density of star clouds within and beyond the galactic center identified the ghostly outline of a small galaxy in the process of colliding with our own. Recognition of this galaxy, the Sagittarius Dwarf Elliptical or SagDEG, led to the realization that M54 is associated with it—this is an extragalactic cluster in the process of being captured by the Milky Way.

R.A. 18h 55m, Dec. -30°29"
Magnitude 8.4
Distance 87,000 light years

OMEGA NEBULA
MESSIER 17

Also known as the Swan Nebula, Messier 17 is best located with binoculars, sweeping down the length of nearby Scutum to the border with Sagittarius looking for a wedge-shaped patch of nebulosity. Through larger telescopes, the nebula takes on a structure similar to the Greek capital Omega (Ω). M17 is one of the brightest and most intense star-forming regions in the Milky Way, and would be a far more spectacular object if it was closer to Earth. As it is, it still offers some rewarding views for stargazers.

R.A. 18h 20m, Dec. -16°11"
Magnitude 6.0
Distance c.5,500 light years

► LAGOON NEBULA
MESSIER 8

Standing out from the Milky Way under the darkest skies, this cloud of delicately glowing gas in western Sagittarius is one of only two star-forming nebulae visible to naked-eye observers from the northern hemisphere. Binoculars will show a bright nucleus at the heart of a glowing cloud spread across the width of three full Moons, while small telescopes will begin to reveal some of its detail including the bisecting stream of dust that gives the nebula its name. A naked-eye star cluster, NGC 6530, overlays the nebula's eastern half and is thought to have emerged in the last 2 million years.

R.A. 18h 04m, Dec. -24°23"
Magnitude 6.0
Distance 5,000 light years

HIGH CONTRAST

This beautiful long-exposure view of the region around Sagittarius and Scorpius captures the blaze of light from countless stars in the Milky Way, contrasting with the spine of dust than runs along the Sagittarius/Carina spiral arm. The bright knots of the Lagoon and Trifid nebulae are on the extreme left, while Antares and the Rho Ophiuchi Complex dominate the right-hand side of the image. The small square marks the location of the Sagittarius A* radio source at the center of the Milky Way, 20,000 light years beyond most structures visible in this picture. Fortunately, invisible radiations such as infrared, radio and X-rays can see through intervening material to lift the veil on our galaxy's central regions.

R.A. 17h 46m, Dec. -29°00"
Magnitude N/A
Distance 26,000 light years

SAGITTARIUS A*

This image from the Chandra X-Ray Observatory is the highest-resolution view of the Milky Way's central region yet obtained. Amid the clouds of reddish lower-energy radiations, lobes of gas emitting powerful X-rays extend on all sides of the Sagittarius A* object. Violent processes are frequently at work in the center of our galaxy, and Hubble Space Telescope observations have revealed the presence of clusters of giant stars surrounding Sagittarius A*, orbiting under the influence of an object that contains 4 million solar masses of material in a region of space smaller than the orbit of Uranus. The fact that there is no visible object coinciding with the radio source and the cluster's center of mass is clinching evidence for a supermassive black hole acting as the gravitational "anchor" for our entire galaxy.

▼ MULTIWAVELENGTH
MILKY WAY

This colorful image combines views of the Milky Way's central region taken by three of NASA's "Great Observatories" satellites. Near-infrared light captured by the Hubble Space Telescope is shown in yellow, a deep-infrared view from the Spitzer Space Telescope appears in red, while data from the Chandra X-Ray Observatory is coded blue. Lifting the veil on the heart our galaxy reveals a landscape of twisted nebulae and dust clouds, shaped by violent stellar winds, supernovae shock waves, and powerful gravitational forces. Sagittarius A*, marking the center of our galaxy, is just to the right of center.

CAPRICORNUS
THE SEA GOAT

This curious zodiac constellation to the east of Sagittarius once marked the spot where the Sun reached its annual southerly extreme. Rather faint and shapeless, it has only one star brighter than magnitude 3.0.

Despite its disadvantages, Capricornus is one of the most ancient constellations—Babylonian and Assyrian astronomers referred to it as the "goat-fish," while other cultures saw it as a more normal goat, ibex, or ox. Greek astronomers, drawing on the Mesopotamian form, saw the constellation as a depiction of the goat-headed god Pan, transforming into a fish to escape from the monstrous Typhon.

The brightest star in this region of the sky is Delta Capricorni. This eclipsing binary system normally shines at magnitude 2.9, but briefly drops by 0.2 magnitudes once every 24.5 hours as its tightly orbiting components pass in front of each other. Alpha Capricorni (Algiedi), meanwhile, is a naked-eye double with two stars of magnitudes 3.6 and 4.2 at wildly different distances (108 and 635 light years). A small telescope will reveal that each of these stars has a fainter binary companion.

CAPTURED CLUSTER Globular cluster Messier 30 is the deep-sky highlight of Capricornus—a dense cloud of stars just over 29,000 light years from Earth, easily detectable through binoculars at magnitude 7.7. Telescopes will reveal that it has undergone a process known as "core collapse," in which large numbers of stars fall into the center of the cluster.

	R.A.	DEC.	TYPE	MAGNITUDE	DISTANCE
Algiedi (α)	20h 18m	-12°31"	Visual double	4.2, 3.6	635, 108 l.y.
Dabih (β)	20h 21m	-14°47"	Multiple star	3.1	330 l.y.
δ Cap	21h 47m	-16°08"	Multiple star	2.8–3.1 (var)	38 l.y.
M30	21h 40m	-23°11"	Globular cluster	7.7	29,400 l.y.
NGC 6907	20h 25m	-24°49"	Spiral galaxy	11.3	140 million l.y.

PISCIS AUSTRINUS AND MICROSCOPIUM

THE SOUTHERN FISH AND THE MICROSCOPE

The southern fish is among the most southerly of the ancient Greek constellations, and is easily identifiable thanks to the presence of the brilliant star Fomalhaut. The more recently invented Microscopium is far more obscure.

In mythological terms, the southern fish was sometimes depicted as parent to the more famous fish in Pisces, drinking from the waters that poured from the cup of Aquarius. Microscopium, in contrast, is one of several more modern constellations invented by French astronomer Nicolas Louis de Lacaille in the 1750s.

Fomalhaut, the brightest star in this region of the sky at magnitude 1.2, is relatively close to Earth at a distance of 25 light years. A relatively young star at around 300 million years old, it is today celebrated for the disk of protoplanetary material (and possible planets) that orbit around it. Beta and Gamma Piscis Austrini, meanwhile, are double stars of middling brightness, which require a small or medium-sized telescope respectively to reveal their fainter companions.

EYE OF GOD This extraordinary image from the Hubble Space Telescope reveals fine structure in the ring of material orbiting Fomalhaut (the star itself lies in the blanked-out area at the center of the picture). The shape of this "debris disk," and in particular the sharp inner edge of the ring, suggests the presence of an unseen planet with at least the mass of Neptune.

	R.A.	DEC.	TYPE	MAGNITUDE	DISTANCE
β PsA	22h 32m	-32°21"	Multiple star	4.3	148 l.y.
γ PsA	22h 53m	-32°53"	Binary star	4.5	220 l.y.
α Mic	20h 50m	-33°47"	Multiple star	4.9	380 l.y.
AU Mic	20h 45m	-31°20"	Planetary system	c.8.6 (var)	32 l.y.

SCULPTOR AND FORNAX

THE SCULPTOR'S WORK-SHOP AND THE FURNACE

These two obscure constellations lie to the south of Cetus, tucked into a bend of the celestial river Eridanus. Both are faint and obscure, but by coincidence each is home to a significant cluster of galaxies.

Fornax and Sculptor are both somewhat typical of the constellations added to southern skies by the French astronomer Nicolas Louis de Lacaille in his 1763 atlas of southern stars. Lacaille was an assiduous observer, cataloging the positions of some 10,000 stars over four years working at the Cape of Good Hope from 1750, but his constellations are mostly faint and hard to identify.

Sculptor lies in the direction of our galaxy's south pole, and offers clear views of intergalactic space. Several nearby galaxies—the closest galaxy group to our own—lie in this direction, while Fornax is host to the far richer and more distant Fornax Cluster. Alpha Fornacis is a binary some 42 light years from Earth, consisting of yellow stars of magnitudes 4.0 and 6.5 that can be split with binoculars or a small telescope. Alpha Sculptoris is a blue giant of magnitude 4.3, 780 light years away.

FORNAX CLUSTER Fornax is home to a rich cluster of galaxies—the second most impressive in our cosmic neighborhood after the Virgo Cluster (see page 104). The Fornax Cluster is centered about 65 million light years from Earth, and is probably linked with another cluster in Eridanus, 20 million light years further away. The core of the cluster occupies a region around 2 degrees across, concentrated around the giant galaxies NGC 1316 and NGC 1365, and is a good target for small telescopes.

	R.A.	DEC.	TYPE	MAGNITUDE	DISTANCE
R Scl	01h 27m	-32°33"	Variable star	5.8–7.7	1,550 l.y.
α For	03h 12m	-28°59"	Binary star	4.0/6.5	42 l.y.
NGC 55	00h 15m	-39°11"	Spiral/Irreg. gal.	8.8	7 million l.y.
NGC 253	00h 48m	-25°17"	Spiral galaxy	7.1	10 million l.y.
NGC 1097	02h 46m	-30°17"	Active galaxy	10.2	45 million l.y.
NGC 1316	03h 23m	-37°13"	Active galaxy	9.4	70 million l.y.

SILVER COIN GALAXY
NGC 253

Midway between Alpha Sculptoris and Beta Ceti lies one of the brightest galaxies in southern hemisphere skies—a nearby spiral that we view from almost edge-on. NGC 253 is the dominant member of the Sculptor Group, the nearest galaxy group beyond our own Local Group. The galaxy's narrow profile concentrates its light and makes it easy to spot with binoculars, while larger telescopes will reveal a mottled ellipse with hints of spiral structure. NGC 253's spiral arms are unusually bright in comparison to its core, indicating that it is currently undergoing a major burst of star formation.

R.A. 0h 48m, Dec. -25°17''
Magnitude 7.1
Distance 10 million light years

BORDERLINE CASE
NGC 55

This peculiar galaxy, displaying characteristics somewhere between barred spiral and irregular galaxies, lies on the southern border of Sculptor, a little to the northwest of Alpha Phoenicis. Although it lies close to the other galaxies of the Sculptor Group, and has traditionally been seen as a member of that group, recent studies suggest it may be an outlying member of our own Local Group, or even an independent galaxy with no gravitational allegiance to either group. Its distortion is almost certainly due to the proximity of the small spiral galaxy NGC 300.

R.A. 0h 15m, Dec. -39°11''
Magnitude 8.8
Distance 7 million light years

RINGED GALAXY
NGC 1097

The bright central regions of this barred spiral in Fornax are visible through a small telescope to the northwest of Beta Fornacis, but larger telescopes are needed to reveal its complete structure. NGC 1097 is currently undergoing a wave of star formation that lights up its spiral arms. The activity may be connected to an ongoing close encounter with the small elliptical galaxy NGC 1097a (at upper left in this picture). One of the galaxy's most unusual features is a perfect ring of star formation surrounding the core, and this is almost certainly linked to the current galactic interaction as well.

R.A. 02h 46m, Dec. -30°17''
Magnitude 10.2
Distance 45 million light years

▶ HUBBLE
ULTRA-DEEP FIELD

Inspired by the success of the original "Hubble Deep Field" observations (see page 46), astronomers turned their attention in 2003 to an even more ambitious project. With a combined exposure time of a million seconds, the Ultra-Deep Field (HUDF) combines visible-light observations with near-infrared data from Hubble's NICMOS camera, revealing objects so distant (and therefore receding rapidly due to the expansion of the Universe) that their light has been Doppler-shifted into the infrared. The result is a view that looks back through time to a mere 400 million years after the Big Bang.

R.A. 03h 33m, Dec. -27°47''
Magnitude < 29.0
Distance Up to 13.3 billion light years

ERIDANUS
THE RIVER

The twisting constellation of the celestial river winds its way back and forth across the sky, trending southward from its origin by Orion's foot, to its end point at brilliant Achernar.

Greek astronomers associated Eridanus with a variety of rivers both real and mythological. However, the ancient constellation ended at the star now designated Theta Eridani. The river was only extended to the south after Renaissance European explorers reported the bright star that lay out of sight from Mediterranean latitudes.

Achernar, whose name derives from the Arabic for "river's end," lies some 143 light years from Earth and is a hot blue-white star shining at magnitude 0.5. It is also one of the fastest-rotating stars in the sky, spinning so rapidly that its equator bulges by 50 percent compared to its poles. Other notable stars include the Sunlike Epsilon Eridani, and the double star Omicron 2 Eridani. A small telescope will reveal that the magnitude-4.4 primary has a companion of magnitude 9.5—the most easily seen white dwarf in the sky.

EPSILON ERIDANI Officially the third-closest naked-eye star system to the Sun, Epsilon Eridani shines at magnitude 3.7 and lies 10.5 light years away. It has the mass of 0.8 Suns and is relatively young at around a billion years old. An excess of infrared radiation was the first hint that Epsilon was surrounded by a disk of potentially planet-forming dust. In 2000, the presence of a giant planet in a seven-year orbit around the star was confirmed, and the system is now known to contain two rocky asteroid belts and possibly at least one more planet.

	R.A.	DEC.	TYPE	MAGNITUDE	DISTANCE
ε Eri	03h 33m	-09°28"	Planetary system	3.7	10.5 l.y.
θ Eri	02h 58m	-40°18"	Binary star	3.2/4.3	120 l.y.
o2 Eri	04h 15m	-07°39"	Multiple star	4.4/9.5/10.5	16.5 l.y.
32 Eri	03h 54m	-02°57"	Binary star	5.0/6.3	290 l.y.
NGC 1535	04h 14m	-12°44"	Planetary nebula	9.6	5,800 l.y.
NGC 1300	02h 20m	-19°25"	Barred spiral gal.	11.4	70 million l.y.

LEPUS

THE HARE

Directly to the south of Orion lies a bow-tie-shaped group of fainter stars representing a hare. Although it is lacking in impressive deep-sky objects, this constellation is home to some intriguing stars.

Lepus is normally associated with a hare being chased by Orion's dogs, and is usually shown crouching unnoticed at the hunter's feet as Orion himself turns to face the charging bull Taurus. Some Arab astronomers, however, saw it as a pair of camels drinking from the sweet waters of the celestial river Eridanus.

Alpha Leporum, or Arneb, is a rare white supergiant of magnitude 2.6, around 1,300 light years from Earth and 13,000 times more luminous than the Sun. Gamma Leporis, meanwhile, is an attractive double star with yellow and orange components of magnitudes 3.6 and 6.2, divisible through binoculars or a small telescope. South of the main star pattern, meanwhile, lies Messier 79, a globular cluster of magnitude 8.6 some 41,000 light years from Earth and best seen with a small telescope.

HIND'S CRIMSON STAR Close to the western boundary of Lepus, binocular observers can find R Leporis, one of the reddest stars in the sky. Also known as Hind's Crimson Star, it lies some 1,500 light years from Earth and is a long-period variable similar to Mira in Cetus (see page 130). It slowly in size and brightness, usually between magnitudes 7.3 and 9.8, with a period of around 420 days. However, on occasion it can become visible to the naked eye, reaching magnitude 5.5.

	R.A.	DEC.	TYPE	MAGNITUDE	DISTANCE
κ **Lep**	05h 33m	-17°49"	Binary star	4.4/7.4	220 l.y.
R Lep	05h 00m	-14°48"	Variable star	5.5–11.7	820 l.y.
M79	05h 25m	-24°33"	Globular cluster	8.6	41,000 l.y.
NGC 2017	05h 39m	-17°51"	Visual cluster	6.4	N/A

CAELUM AND COLUMBA

THE CHISEL AND THE DOVE

Two small constellations lie to the south of Lepus, between the equatorial stars of Orion and the equally prominent southern stars of Vela and Carina. These form the relatively bright constellation of the dove and the more obscure chisel.

Columba owes its origins to Dutch astronomer and theologian Petrus Plancius, who inserted several biblically inspired constellations into a celestial chart published in 1592—he intended it to represent the dove sent out by Noah from the Ark. Caelum, meanwhile, is an invention of French astronomer Nicolas Louis de Lacaille from the 1750s.

Alpha Columbae, or Phact, is a fast-spinning blue-white star of average magnitude 2.6 around 530 light years from Earth. As it spins, it flings out material from around its equator, creating a series of rings that cause its brightness to fluctuate slightly. Another interesting star is Mu Columbae, a hot blue stellar runaway of magnitude 5.1, around 1,300 light years from Earth. Its fast-moving track through space, like that of AE Aurigae (see page 40), can be traced back to the Orion Nebula.

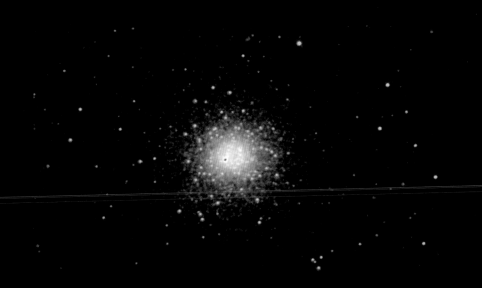

MIXED GLOBULAR NGC 1851, in the southwestern corner of Columba, is one of the sky's more distant globular clusters, at 39,500 light years from Earth. Despite this, it is relatively bright at magnitude 7.3 and can usually be spotted through good binoculars. NGC 1851 displays several unusual features that suggest it may have formed from the merger of two separate globular clusters at some time in its history.

	R.A.	DEC.	TYPE	MAGNITUDE	DISTANCE
Phact (α Col)	05h 40m	-34°04"	Variable star	c.2.6 (var)	530 l.y.
γ1 Cae	05h 04m	-35°29"	Binary star	4.6/8.1	185 l.y.
NGC 1851	05h 14m	-40°03"	Globular cluster	7.3	39,500 l.y.

PUPPIS
THE STERN OF ARGO

The northernmost part of Argo, the great celestial ship, rises into equatorial skies to the south and east of Canis Major. Crossed by a broad swathe of the Milky Way, Puppis is rich in dense star clouds and clusters.

Puppis marks the stern of the mighty ship Argo. Ancient Greek stargazers recognized this huge constellation sailing across their southern horizon during the northern hemisphere spring, but later astronomers split it into three separate parts, the stern, keel (Carina), and sail (Vela).

One peculiar consequence of Argo's dissolution is that the three modern constellations share one set of Greek "Bayer letters" between them—hence Naos, this constellation's brightest star, is designated Zeta Puppis. This extremely hot blue star of magnitude 2.2 is just 4 million years old and lies roughly 1,100 light years away. Another highlight is L Puppis, a line-of-sight double star easily separated with binoculars. The northern component, blue-white L1 Puppis, shines at a steady magnitude 4.9, but its southern companion, the reddish L2, varies between magnitude 2.6 and 6.2 in a 141-day cycle.

CAULDRON OF STARBIRTH The seething and bubbling gases revealed within nebula NGC 2467 by this Hubble Space Telescope image have been compared to a witch's cauldron. The nebula lies close to Omicron Puppis and forms the background to a star cluster of magnitude 7.1 that can be located with binoculars. Despite appearances, the cluster's stars are thought to lie some way in front of the nebula.

	R.A.	DEC.	TYPE	MAGNITUDE	DISTANCE
Asmidiske (ξ)	07h 49m	-24°52"	Multiple star	3.3	1,350 l.y.
L Pup	07h 44m	-28°57"	Visual double	4.9, 2.6–6.2 (var)	181, 198 l.y.
M46	07h 42m	-14°49"	Open cluster	6.0	5,400 l.y.
M47	07h 37m	-14°30"	Open cluster	4.2	1,600 l.y.
M93	07h 45m	-23°52"	Open cluster	6.0	3,600 l.y.
NGC 2451	07h 45m	-37°58"	Open cluster	2.8	850 l.y.

ANTLIA AND PYXIS

THE AIR PUMP AND THE COMPASS

These two obscure constellations of mid-southern skies are best found by looking in the empty region west of Centaurus and east of Puppis, northward of the bright stars of Vela.

Pyxis and Antlia are both 18th-century inventions of the French astronomer Nicolas de Lacaille, and are typically unconvincing as star patterns. Like most of Lacaille's constellations, they represent scientific instruments of the time—in this case a magnetic compass and a pneumatic air pump.

Zeta Antliae is a deceptive double—binoculars reveal a pair of white stars of magnitudes 5.8 and 5.9, but even though they are at roughly the same distance (around 370 light years), they are not bound in orbit around each other, and so are technically a line-of-sight double. Small telescopes show that the western component (Zeta 1) is a true binary of magnitudes 6.2 and 7.0. T Pyxidis, meanwhile, is a recurrent nova—a normally obscure system of magnitude 13.8, some 6,000 light years from Earth, that flares into binocular or even naked-eye visibility every two or three decades (most recently in 1986).

TILTED SPIRAL NGC 2997 is a bright spiral galaxy to the southwest of Theta Antliae, presented at an angle of roughly 45 degrees. It lies around 40 million light years from Earth and forms the gravitational anchor for its own small galaxy group. With a diameter roughly one-third that of the full Moon and a magnitude of 10.1, it is a good challenge for locating with smaller telescopes, though much larger instruments are needed to bring out its detail.

	R.A.	DEC.	TYPE	MAGNITUDE	DISTANCE
δ Ant	10h 30m	-30°36″	Binary star	5.6/9.7	480 l.y.
ζ, Ant	09h 31m	-31°53″	Binary star	6.2/7.0	370 l.y.
T Pyx	09h 05m	-32°23″	Recurrent nova	6.0–13.8	6,000 l.y.

VELA
THE SAIL OF ARGO

The irregular octagonal shape of this constellation is not the most obvious of patterns in the sky, but it is easily identifiable thanks to its brightness and location between Centaurus and Crux to the east, and Carina to the west.

Vela marks the sail of the ship Argo, an ungainly constellation that has since been broken into three segments. In mythology, Argo was built for Jason by the great shipwright Argus, and placed under the protection of Hera, the Queen of the Gods.

The constellation's brightest star, Gamma Velorum, or Regor, is a complex multiple star of magnitude 1.8. Binoculars reveal one companion star of magnitude 4.3, while small telescopes will show two more of magnitudes 8.5 and 9.4. Studies of the spectrum of Gamma 1's light reveal that it is also a binary star with components 10 and 30 times the mass of the Sun. Delta Velorum is another multiple, its most obvious components a white star of magnitude 2.0, and a yellow one of magnitude 5.1. Another attractive object is IC 2391, a naked-eye open cluster centered on Omicron Velorum.

VELA SUPERNOVA REMNANT Northeast of Gamma Velorum, a network of faintly glowing gaseous tendrils marks the outline of the Vela Supernova Remnant (SNR)—an expanding shockwave of hot gas generated by a stellar explosion around 11,000 years ago. For visual observers, the Vela SNR is frustratingly faint, and best seen through long-exposure photographs.

	R.A.	DEC.	TYPE	MAGNITUDE	DISTANCE
γ **Vel**	08h 10m	-47°20"	Multiple star	1.8/4.3/8.5/9.4	1,200 l.y.
δ **Vel**	08h 45m	-54°43"	Multiple star	2.0/5.1	80 l.y.
NGC 3132	10h 07m	-40°26"	Planetary nebula	9.9	2,000 l.y.
NGC 3201	10h 18m	-46°25"	Globular cluster	6.8	15,000 l.y.
IC 2391	08h 40m	-53°04"	Open cluster	2.5	580 l.y.

VELA INSIDE VIEW

HH 47
Herbig Haro Objects are twin-lobed nebulae found on either side of young stars and often connected to them by narrow jets. They are formed when a rapidly spinning newborn star flings off excess material along its axis of rotation. As these high-speed jets stream out into space, they may encounter interstellar gas and dust, creating a glowing shock wave. Herbig Haro 47 (HH 47) is perhaps the best known and most studied example of these intriguing objects, with jets that display numerous complex knots throughout their length. The Hubble Space Telescope image above shows the shock front on one side of the nebula.

EIGHT-BURST NEBULA NGC 3132
This bright planetary nebula lies on Vela's northern boundary, northwest of the magnitude-3.8 triple star p Velorum. Seen through small telescopes, it forms a disk of light somewhat larger than Jupiter, with a magnitude-10.0 star sitting neatly at its center. However, appearances can be deceptive—the bright star has a binary companion of magnitude 16.0 (beyond the reach of most amateur telescopes), and it is this star that is actually responsible for the surrounding nebula—the brighter star is merely an innocent bystander.

VELA PULSAR PSR B0833-45
From the center of the Vela Supernova Remnant, a rapidly spinning neutron star sends a beam of radio waves, X-rays and gamma rays toward Earth every 89 milliseconds. This "pulsar" signal is created by a powerful magnetic field around the stellar corpse, which channels radiation into two narrow beams that sweep around the sky. This X-ray view of the pulsar reveals that it has a second pair of beams—jets emitted along the pulsar's axis of rotation, which spray energized particles into the surrounding nebula like an out-of-control firehose.

▶ PENCIL NEBULA NGC 2736
The brightest part of the Vela Supernova Remnant at visible wavelengths, the Pencil Nebula marks an expanding shockwave traveling through space at roughly 400,000 mph (650,000 km per hour). Where particles in the shock front collide with those in the interstellar medium, they are excited, creating a glowing wall three-quarters of a light year long, moving across space with awesome speed. The Pencil was discovered by British astronomer John Herschel, working at the Cape of Good Hope in 1835, but was not linked to the larger supernova remnant until much more recently.

R.A. 08h 26m, Dec. -59°03"
Magnitude: c.10.0 (var)
Distance 1,470 light years

R.A. 10h 07m, Dec. -40°26"
Magnitude 9.9
Distance 2,000 light years

R.A. 08h 35m, Dec. -45°11"
Magnitude 23.6
Distance 815 light years

R.A. 09h 00m, Dec. -45°54"
Magnitude 12.0
Distance 815 light years

MYSTIC MOUNTAINS This bizarre and tortured landscape is formed by a variety of star-forming processes at work in the Carina Nebula, some 7,500 light years from Earth. The dark outcrops are similar to the "pillars of creation" found in other nebulae, and are slowly being eroded by radiation from nearby brilliant stars.

	R.A.	DEC.	TYPE	MAGNITUDE	DISTANCE
η Car	10h 45m	-59°41″	Binary star	c.4.6 (var)	7,500 l.y.
υ Car	09h 47m	-65°04″	Binary star	3.1/6.3	1,400 l.y.
S Car	10h 09m	-61°33″	Variable star	4.4–9.9	1,320 l.y.
NGC 2516	07h 58m	-60°52″	Open cluster	3.8	1,300 l.y.
NGC 3372	10h 44m	-59°52″	Diffuse nebula	1.0	7,500 l.y.
NGC 3532	11h 06m	-58°40″	Open cluster	3.0	1,300 l.y.
IC 2602	10h 43m	-64°24″	Open cluster	1.9	480 l.y.

◀ STAR ON THE BRINK ETA (η) CARINAE

A wide view of the Carina Nebula reveals the bulbous "Hounculus Nebula" around Eta Carinae at its heart. One of the most remarkable stars in the sky, Eta currently shines at an unspectacular magnitude 4.6, but is an unpredictable variable: during an outburst that peaked around 1843, it briefly became the second-brightest star in the entire sky. Eta consists of a pair of blue supergiants, each with the mass of around 60 to 80 Suns and pumping out hundreds of thousands of times more energy. Both stars will end their lives as supernovae in the astronomically near future.

R.A. 10h 45m, Dec. -59°41"
Magnitude c.4.6 (var)
Distance 7,500 light years

MASSIVE CLUSTER NGC 3603

When British astronomer John Herschel discovered this cluster while working in South Africa in 1834, he at first thought that it might be globular in nature. In reality, however, NGC 3603 is a particularly dense open cluster, perhaps just a million years old and containing the greatest concentration of heavyweight stars known in the Milky Way galaxy. Though the cluster is surrounded by a dense cloud of gas and dust, fierce radiation and strong stellar winds from its stars have cleared out its surroundings to leave it beautifully framed in empty sky.

R.A. 11h 15m, Dec. -61°16"
Magnitude 9.1
Distance 20,000 light years

BULLET CLUSTER 1E-0657-558

In an empty corner of western Carina, powerful telescopes reveal this distant cluster, about 3.7 billion light years from Earth. The Bullet formed during a dramatic collision between galaxy clusters that began around 150 million years ago. This image shows the distribution of mass (blue) and X-ray emitting gas (pink) between the colliding clusters—it reveals that while the clouds of "cluster gas" have collided head-on, both the galaxies and the mysterious "dark matter," which provides much of the cluster's mass, have passed through each other almost unaffected.

R.A. 06h 59m, Dec. -55°57"
Magnitude 14.2 and fainter
Distance 3.7 billion light years

HERSCHEL'S VIEW

This vivid portrait of the Carina Nebula is a product of the European Space Agency's Herschel Space Observatory, an orbiting telescope designed for investigating the far reaches of the infrared spectrum and some of the coldest objects in the Universe. It reveals that the visible parts of the nebula are surrounded by enormous and previously unsuspected reserves of invisible gas, giving the entire nebula an estimated mass of 350,000–900,000 Suns. Despite being invisible to other instruments, this gas is clearly not immune to the forces that shape the rest of the nebula—there are many places where it has been carved into pillars or hollowed out in bubbles by the influence of the superhot stars being created within it.

R.A. 10h 45m, Dec. -59°52"
Magnitude 1.0
Distance 7,500 light years

HUBBLE FANTASY

This segment from a mosaic constructed using 48 separate Hubble Space Telescope images covers an area of the Carina Nebula roughly a dozen light years wide—the entire nebula is more than 200 light years across. Opaque loops of dusty gas on the left of the image form part of the top of the Keyhole Nebula, a dark nebula that is one of the complex's most prominent features in amateur telescopes. In the center, meanwhile, are several isolated "Bok globules." These dense pockets of gas and dust, set adrift from their origin in "pillars of creation" may each contain an incubating star or star system.

▶ INFRARED LANDSCAPE
This beautiful near-infrared view
of the Carina Nebula was captured
by the HAWK-I camera on the
European Southern Observatory's
Very Large Telescope. It reveals
hitherto unsuspected arcs of
gas close to Eta Carinae (lower
left), dark star-forming regions,
previously undiscovered cooler
stars shown in yellow and a
spectacular central star cluster
called Trumpler 14. This cluster
is home to the most recent stars
produced by the nebula, at around
half a million years old, and may
contain as many as 2,000 stars in
a region just six light years across.

CRUX

THE SOUTHERN CROSS

The smallest constellation in the sky is also one of the most distinctive, a compact group of four bright stars nestling beneath the body of Centaurus.

The origins of this renowned constellation are surprisingly uncertain, but Crux appears to have been invented in the early 16th century. Earlier astronomers knew of its stars, but considered them to be part of Centaurus.

Three of the key stars (Alpha, Beta, and Delta) are blue-white giants at 320–350 light years from Earth. Like many other bright stars in this part of the sky, they belong to the Scorpius-Centaurus OB Association and are around 10–20 million years old. Alpha Crucis (Acrux) is a bright binary pair, whose members, shining at magnitudes 1.3 and 1.7 are easily divisible with the smallest telescope.

The fourth star of the Cross, Gamma Crucis, is a red giant of magnitude 1.6, just 88 light years away. While bright Milky Way starfields form the backdrop to much of the constellation, a prominent dark nebula, known as the Coalsack, occupies its southeast quadrant.

THE JEWEL BOX Visible to the naked eye as a fuzzy star of magnitude 4.2, the open cluster NGC 4755 is a highlight of southern skies. It lies just to the east of Beta Crucis, with Kappa Crucis as its brightest individual star at magnitude 5.9. Binoculars or a telescope on low magnification reveal a spectacular cloud of several dozen stars, mostly blue-white but with a brilliant red giant of magnitude 7.6 forming a pleasing contrast. The Jewel Box lies far beyond the brighter stars of Crux, around 7,600 light years away, and is thought to be just 7 million years old.

	R.A.	DEC.	TYPE	MAGNITUDE	DISTANCE
Acrux (α)	12h 27m	-63°06"	Multiple star	1.3/1.7	320 l.y.
Mimosa (β)	12h 48m	-59°41"	Variable star	1.3	350 l.y.
NGC 4755	12h 54m	-60°20"	Open cluster	4.2	7,600 l.y.
Coalsack	12h 52m	-63°18"	Dark nebula	N/A	2,000 l.y.

MUSCA
THE FLY

This small, bright but somewhat jumbled constellation is easily identified thanks to its position in the Milky Way to the south of Crux. Unsurprisingly, given this location, it contains several interesting objects.

Musca was invented in the late 16th century by Dutch astronomer Petrus Plancius, based on reports of the southern sky brought back by Dutch navigators. Originally known as Apis, the bee, it was reimagined as a fly by Nicolas Louis de Lacaille in 1752—in part to avoid confusion with Apus (see page 226).

Alpha Muscae is a massive blue-white star of magnitude 2.7, around 315 light years from Earth. The slightly more distant Beta Muscae, at magnitude 3.0, is a binary whose evenly matched components can be split with a small telescope and high magnification. Theta Muscae is another binary, with components of magnitudes 5.7 and 7.3. Both stars are hot and blue stellar heavyweights—the fainter component is a rare Wolf-Rayet star—one that has blown away significant amounts of material in stellar winds during its lifetime, revealing even hotter inner layers.

THE HOURGLASS NEBULA Also known as MyCn 18, this planetary nebula in Musca's northwestern corner lies around 8,000 light years from Earth. With an apparent magnitude of 13.0, it is unfortunately beyond most amateur instruments, and its true structure was not appreciated until 1996, when the Hubble Space Telescope captured this remarkable image.

	R.A.	DEC.	TYPE	MAGNITUDE	DISTANCE
β Mus	12h 46m	-68°06"	Binary star	3.5/4.0	340 l.y.
θ Mus	13h 08m	-65°18"	Binary star	5.7/7.3	c.10,000 l.y.
NGC 4833	13h 00m	-70°53"	Globular cluster	7.8	21,500 l.y.

CIRCINUS AND TRIANGULUM AUSTRALE

THE COMPASSES AND THE SOUTHERN TRIANGLE

Two celestial triangles—one long and thin, the other broad and almost equilateral— lie to the east of brilliant Alpha and Beta Centauri in the deep southern skies.

The broad southern triangle is the older of these two constellations—first recorded in Johann Bayer's *Uranometria* star atlas of 1603, and probably invented by Dutch navigator Pieter Dirkszoon Keyser a few years before. Dating to the 1750s, the narrower triangle of the compasses is an invention of Nicolas Louis de Lacaille.

Alpha Trianguli Australis is an orange giant of magnitude 1.9, some 415 light years from Earth, while Alpha Circini is a white star of magnitude 3.2, one of a class of "Delta Scuti" variables that changes its brightness almost imperceptibly with a period of just a few minutes. Gamma Circini is an attractive double with a blue-white component of magnitude 5.1 and a companion of magnitude 5.5 that sometimes appears yellow in contrast, but is in fact a pure white. NGC 6025 in Triangulum Australe is a tight open star cluster of magnitude 5.1, resolvable through binoculars.

CIRCINUS GALAXY Despite its relative closeness to Earth at just 13 million light years, the dense Milky Way star clouds that run behind Circinus kept this curious galaxy hidden and undiscovered until the 1970s. Now cataloged as ESO 97-G13, it is the closest example of an active galaxy with a bright nucleus that pours out large amounts of radiation from the region around its central supermassive black hole.

	R.A.	DEC.	TYPE	MAGNITUDE	DISTANCE
α Cir	14h 43m	-64°59"	Binary star	3.2	53 l.y.
γ Cir	15h 23m	-59°19"	Binary star	5.1/5.5	510 l.y.
θ Cir	14h 57m	-62°47"	Variable star	4.8–5.7	830 l.y.
NGC 6025	16h 04m	-60°30"	Open cluster	5.1	2,700 l.y.

NORMA
AND ARA

These two constellations are embedded in a rich stream of the Milky Way to the south of Scorpius. Their stars are far from impressive, but they are rich in interesting deep-sky objects such as star clusters.

Despite its apparent obscurity, Ara is one of the oldest constellations—recognized as a sacrificial altar by the earliest Mesopotamian stargazers. Over a thousand years later, the Greeks saw it as the altar of the Gods of Olympus. Norma, in contrast, was invented by Nicolas de Lacaille in the 1750s.

Gamma Normae is a line-of-sight double consisting of a magnitude-5.0 yellow supergiant some 1,450 light years from Earth, and a magnitude-4.0 yellow giant just 127 light years away. The pair can usually be separated with the naked eye. S Normae, at the heart of the NGC 6087 open star cluster, is another yellow supergiant and a variable star similar to Delta Cephei (see page 34), with a 9.8-day cycle that sees its magnitude vary between 6.1 and 6.8. NGC 6397 in Ara, meanwhile, is one of the nearest globular clusters at 7,200 light years, and easily spotted through binoculars.

GREAT ATTRACTOR Hidden behind the Milky Way in Norma lies a dense galaxy cluster catalogued as Abell 3627 but generally known as the Norma Cluster. It is relatively close to Earth at 220 million light years, but best observed using X-rays (as shown here) to pierce the intervening star clouds. The cluster's location coincides with a mysterious object known as the Great Attractor—a dense concentration of mass that draws everything else in the local Universe towards it.

	R.A.	DEC.	TYPE	MAGNITUDE	DISTANCE
γ Nor	16h 20m	-50°09"	Visual double	5.0, 4.0	1,450, 127 l.y.
μ Arae	17h 44m	-51°50"	Planetary system	5.2	50 l.y.
NGC 6087	16h 18m	-57°56"	Open cluster	5.6	2,700 l.y.
NGC 6193	16h 41m	-48°44"	Open cluster	5.2	4,300 l.y.
NGC 6397	17h 41m	-53°40"	Globular cluster	6.7	7,200 l.y.

CORONA AUSTRALIS AND TELESCOPIUM

THE SOUTHERN CROWN AND THE TELESCOPE

Directly to the south of Sagittarius lies an easily identified arc of stars forming the Southern Crown, a counterpart to the northern Corona Borealis. South of this lies the far more obscure constellation of the Telescope.

The Southern Crown is an ancient constellation, rising just high enough to be visible from Greek latitudes. It is often associated with Bacchus, the god of wine, or sometimes with the nearby centaur, Sagittarius. Telescopium is one of Nicolas Louis de Lacaille's 18th-century additions to the sky, and possibly the most obscure constellation of all.

Gamma Coronae Australis is an attractive binary that can be separated with small telescopes to reveal near-twin yellow stars of magnitudes 4.8 and 5.1. Eta and Kappa Coronae Australis, meanwhile, are both line-of-sight doubles, easily separated with binoculars. For amateur astronomers, the constellation's deep-sky highlight is the dense and bright globular cluster NGC 6541, 23,000 light years away and shining at magnitude 6.3.

R CORONAE AUSTRALIS The region around the young and unpredictable variable star R Coronae Australis, 420 light years away, is home to a complex of dark, reflection and emission nebulae. As one of the nearest star-forming regions to Earth, it offers a rare glimpse of the environment in which less massive, Sunlike stars are born. Such stars do not release the high-energy radiation that illuminates more spectacular star-forming regions, so the nebula's colors are far more restrained.

	R.A.	DEC.	TYPE	MAGNITUDE	DISTANCE
γ CrA	19h 06m	-37°04"	Binary star	4.8/5.1	58 l.y.
δ Tel	18h 32m	-45°55"	Visual double	4.9, 5.1	800, 1080 l.y.
NGC 6541	18h 08m	-43°43"	Globular cluster	6.3	23,000 l.y.

PAVO

THE PEACOCK

South of Corona Australis and the obscure Telescopium, the constellation of the Peacock is best located through the presence of the bright star, Peacock.

While most star names can be traced back to ancient Greek, Latin, or Arabic roots, the origins of Alpha Pavonis's proper name are rather more recent—it was invented when the star was chose for inclusion in navigation manuals by Britain's Royal Air Force in the 1930s. Pavo itself, meanwhile, is one of several "southern birds" added to the sky by Petrus Plancius in 1597, based on the reports of Dutch navigators.

At magnitude 1.9, bright-blue Peacock is a spectroscopic binary—a star that we know is a pair thanks to the analysis of its light, but which cannot be visually separated. Kappa Pavonis, meanwhile, is a pulsating Cepheid variable—a yellow supergiant 490 light years away that fluctuates between magnitudes 3.9 and 4.8 in a 9.1-day cycle. Delta Pavonis, 20 light years away and shining at magnitude 3.6, is a star of similar mass to the Sun, but somewhat older and already in the process of evolving into a red giant.

BRIGHT GLOBULAR NGC 6752 in northern Pavo is the third-brightest globular cluster in the entire sky, easily visible with binoculars at magnitude 5.4, and about two-thirds the diameter of the full Moon. It lies 13,000 light years from Earth and contains more than 100,000 stars in a region of space roughly 100 light years across. This Hubble Space Telescope image covers the central ten light years, showing that the density of stars increases exponentially towards the cluster's core.

	R.A.	DEC.	TYPE	MAGNITUDE	DISTANCE
Peacock (α)	20h 26m	-56°44"	Multiple star	1.9	183 l.y.
κ Pav	18h 57m	-67°14"	Variable star	3.9–4.8	490 l.y.
NGC 6752	19h 11m	-59°59"	Globular cluster	5.4	13,000 l.y.
NGC 6744	19h 10m	-63°51"	Spiral galaxy	9.1	25 million l.y.

GRUS AND PHOENIX

THE CRANE
AND THE PHOENIX

Two birds, one real and one mythological, straddle the line between the bright stars Fomalhaut and Achernar in far southern skies. Both are fairly easy to identify—the Phoenix from its outstretched wings, and the Crane from its elongated neck.

These "southern birds" were first recorded by Johann Bayer in his *Uranometria* star atlas of 1603, but, like those attributed to Petrus Plancius, they started out as inventions of Pieter Dirkszoon Keyser and Frederick de Houtman, a pair of Dutch navigators who kept detailed astronomical records while trading in the East Indies during the 1590s.

    Zeta Phoenicis is a triple star system: it consists of an eclipsing binary that normally shines at magnitude 3.9 but drops by half a magnitude during eclipses every 1.67 days, and a faint, close companion of magnitude 6.9. Beta Gruis, meanwhile, is a pulsating red giant that varies irregularly between magnitudes 2.0 and 2.3. Unfortunately, these constellations have no significant deep-sky objects suitable for smaller telescopes.

ROBERT'S QUARTET Although at magnitudes around 14.0 these galaxies are too faint to be seen through most amateur instruments, this attractive quartet in Phoenix is still one of the sky's more famous compact galaxy groups. With an ungainly catalog number of AM 0018-485, these four galaxies lie around 160 million light years from Earth. They are contained in an area of sky with one-tenth the diameter of the full Moon, corresponding to a region around 75,000 light years across.

	R.A.	DEC.	TYPE	MAGNITUDE	DISTANCE
Ankaa (α Phe)	00h 26m	-42°18"	Binary star	2.4	77 l.y.
ζ **Phe**	01h 08m	-55°14"	Multiple star	3.9–4.4/6.9	280 l.y.
β **Gru**	22h 43m	-46°53"	Variable star	2.0–2.3	170 l.y.

TUCANA
AND INDUS

THE TOUCAN
AND THE INDIAN

While both of these constellations are fairly faint, Tucana is home to some stunning deep-sky objects including the Small Magellanic Cloud. It can be found easily by looking southwest of bright Achernar in Eridanus. Shapeless Indus lies further to the west, but has little to recommend it.

Both of these constellations were introduced to European astronomers through the reports of Dutch navigators Pieter Dirkszoon Keyser and Frederick de Houtman, who traveled widely in southeast Asia during the 1590s. There is some evidence that they may be based on earlier star patterns already recognized in the region around the East Indies.

Alpha Indi is an orange giant of magnitude 3.1, 100 light years from Earth. Medium-sized telescopes reveal two red-dwarf companions of magnitudes 11.9 and 12.5. Epsilon Indi is a nearby star, just 11.8 light years away and shining at magnitude 4.7. It is a little less massive and less luminous than the Sun, and has its own pair of faint companions—failed stars or "brown dwarfs" that were only discovered in 2003.

OVERLOOKED GLOBULAR Tucana is home to two famous wonders of the heavens—the Small Magellanic Cloud and the bright globular cluster 47 Tucanae (NGC 104). As a result, another globular within Tucana's boundaries tends to be overlooked. At 30,000 light years from Earth, NGC 362 is far more distant than 47 Tucanae, and shines at magnitude 6.4, making it an easy target for binoculars or a small telescope. Larger instruments will start to resolve the cluster's individual stars.

	R.A.	DEC.	TYPE	MAGNITUDE	DISTANCE
α Ind	20h 38m	-47°17"	Multiple star	3.1/11.9/12.5	100 l.y.
ε Ind	22h 03m	-56°47"	Multiple star	4.7	11.8 l.y.
α Tuc	22h 19m	-60°16"	Binary star	2.9	199 l.y.
β Tuc	00h 32m	-62°57"	Visual double	4.3, 5.1	144, 151 l.y.
κ Tuc	01h 15m	-68°53"	Multiple star	4.3	67 l.y.
NGC 104	00h 24m	-72°05"	Globular cluster	4.9	16,700 l.y.
SMC	00h 53m	-72°50"	Irregular galaxy	2.3	210,000 l.y.
NGC 362	01h 03m	-70°51"	Globular cluster	6.4	30,000 l.y.

TUCANA 47 TUCANAE NGC 104

CELESTIAL GLITTERBALL

Easily visible to the naked eye at magnitude 4.9, globular cluster 47 Tucanae looks like a slightly fuzzy star on the edge of the Small Magellanic Cloud. It was French astronomer Nicolas Louis de Lacaille who first noted its non-stellar appearance in 1751, but rather like Omega Centauri's Greek "Bayer letter," the "Bode number" designating the object as a star has stuck. Binoculars will transform the cluster into a ball of light roughly the size of the full Moon, but this cluster is so densely packed that even its outer edges can only be resolved with a medium-sized telescope.

R.A. 00h 24m, Dec. -72°05"
Magnitude 4.9
Distance 16,700 light years.

CENTRAL REGIONS

This image from the European Southern Observatory's Very Large Telescope zooms in on the central portion of 47 Tucanae, resolving detail almost all the way to the core. At around 16,700 light years away, the cluster is one of the closer globulars to Earth, but it is also a genuinely impressive object, with a total mass equivalent to a million Suns packed into a region 120 light years across. In the center of the cluster, stars are separated by light days rather than light years.

PLANET SURVEY

This Hubble Space Telescope view captures a field of 35,000 stars within 47 Tucanae, studied in 1999 in an effort to search for extrasolar planets in this crowded environment. The telescope stared at these stars for eight days, looking for telltale dips in their brightness caused by planetary transits (in which a planet crosses the face of its star and blocks out some of its light). Although such events are rare, this "transit technique" has been used successfully elsewhere, and statistically, astronomers expected to detect around 17 transits. The fact that the survey found none suggests that conditions inside globular clusters either prevent the birth of planets, or strip them away from their parent stars after formation.

▶ STARS IN MOTION

This extreme close-up of 47 Tucanae's core is just one in a series of Hubble Space Telescope images that astronomers used to track the motion of more than 15,000 stars within the cluster. By comparing the speeds of the stars with their masses (calculated from their brightness and evolutionary state), they were able to map the complex dynamics at work inside clusters for the first time. Frequent close encounters within 47 Tucanae exchange momentum between stars, ultimately leading heavier stars to slow down in their orbits and "sink" towards the core of the cluster, while accelerating lightweight stars and boosting their orbits towards the cluster's outer limits.

TUCANA THE SMALL MAGELLANIC CLOUD

SATELLITE GALAXY

Clearly visible to the naked eye in southeastern Tucana, the Small Magellanic Cloud (SMC) resembles a small, isolated fragment of the Milky Way, around seven times the diameter of the full Moon. Binoculars and small telescopes will start to reveal a wealth of detail including star clusters and nebulae, while the constellation presents almost limitless opportunities for larger instruments. Compared to its "big brother," the Large Magellanic

smaller (at around 10,000 light years across) and further away at 210,000 light years. The clusters take their name from Portuguese explorer Ferdinand Magellan, who was the first European to report them during his round-the-world voyage of 1519–21.

R.A. 00h 53m, Dec. -72°50˝
Magnitude 2.3
Distance 210,000 light years

STELLAR NURSERY
N66 AND NGC 346

Both the Large and Small Magellanic Clouds are classed as irregular galaxies, rich in gas and dust that supply the raw materials for ongoing star formation. N66 is one of many beautiful starbirth nebulae in the SMC, a glowing cocoon illuminated from within by the intense radiation of a recently formed star cluster known as NGC 346. At magnitude 10.3, the cluster can be located through

long photographic exposures are needed to reveal the surrounding nebulosity. NGC 346 contains many of the SMC's most massive stars—the shockfront generated by their powerful stellar winds is clearly visible at the top and left of this Hubble Space Telescope image.

R.A. 00h 59m, Dec. -72°10˝
Magnitude 10.3
Distance 210,000 light years

► HIDDEN TREASURE
N90 AND NGC 602

This stunning Hubble image showcases one of the SMC's most beautiful star-forming regions, known as N90. This cavern-like nebula is shaped by radiation and stellar winds blasting out from NGC 602, the newborn cluster of brilliant stars at its heart. As the walls of the nebula fight their losing battle against the forces of erosion, denser outcrops, similar to the "pillars of creation" found in the Eagle Nebula (see page 114) stand out like stalagmites and stalactites. Near the bottom of the picture, a distant cluster of galaxies, millions of light years beyond N90, emerges through the hazy gas.

R.A. 01h 29m, Dec. -73°34″
Magnitude 13.1
Distance 210,000 light years

HOROLOGIUM
AND RETICULUM
THE CLOCK AND THE RETICLE

Tucked beneath the southern reach of
the celestial river Eridanus, these two
constellations are faint and obscure.
The Clock (intended to show the form of
a swinging pendulum) is particularly hard
to spot, but the tight diamond of the Reticle
(or Crosshair) is at least a little clearer.

Both of these rather threadbare constellations stand
mainly as a testimony to the imagination and
determination of their inventor, French astronomer
Nicolas Louis de Lacaille, in his 18th-century effort
to fill every last gap in the southern sky.

Alpha Horologii, marking the pivot of the clock's
pendulum, is an orange giant of magnitude 3.9,
around 117 light years from Earth. Alpha Reticuli is a
binary consisting of a yellow giant of magnitude 3.4,
and a faint red dwarf of magnitude 12.0, visible with
a medium-sized telescope. Perhaps the most
famous star in either constellation, however, is Zeta
Reticuli, a binary pair of Sunlike stars 39 light years
away, shining at magnitudes 5.5 and 5.2 and divisible
with binoculars. Zeta is well known as the supposed
home of the "gray" aliens beloved of UFO
enthusiasts.

NGC 1559 This barred spiral galaxy in Reticulum is unfortunately beyond the
reach of most amateur telescopes, but views from instruments such as the Hubble
Space Telescope reveal it to be a beautiful and active galaxy with a bright core and
well-defined spiral arms alive with star formation.

	R.A.	DEC.	TYPE	MAGNITUDE	DISTANCE
α Ret	04h 14m	-62°28"	Binary star	3.4	165 l.y.
ζ Ret	03h 18m	-62°31"	Binary star	5.2/5.5	39 l.y.
R Hor	02h 54m	-49°53"	Variable star	4.0–14.0	1,000 l.y.
NGC 1261	03h 12m	-55°13"	Globular cluster	8.4	54,000 l.y.
NGC 1512	04h 04m	-43°21"	Spiral galaxy	11.1	30 million l.y.

PICTOR AND DORADO

THE PAINTER'S EASEL AND THE DORADO

These two faint and indistinct constellations lie south of a line joining the bright stars Canopus in Carina and Achernar in Eridanus. While Dorado is home to the spectacular Large Magellanic Cloud, Pictor has no such impressive attractions.

Dorado is another invention of the Dutch navigators Keyser and de Houtman, added to the sky around 1600. Although its name translates as "goldfish," it was probably intended to represent the Hawaiian *mahi-mahi*. Pictor, meanwhile, is another of Lacaille's southern constellations from the 1750s.

Beta Doradus is one of the brightest variable stars in the sky—a pulsating yellow supergiant that varies in brightness between magnitudes 3.5 and 4.1 in a 9.9-day cycle. This Cepheid variable lies 1,040 light years from Earth, and its variations can easily be tracked with reference to other nearby stars. Delta Pictoris, meanwhile, is an eclipsing binary that dips from magnitude 4.7 to 4.9 every 40 hours as one of its hot blue-white stars passes in front of the other. Its variations are just detectable to the human eye.

BETA PICTORIS Pictor's most famous object is an unassuming white star of magnitude 3.9, 63 light years from Earth. It owes its fame to a broad disk of gas and dust that swirls around it, extending to more than 40 times the width of Neptune's orbit. This "protoplanetary disk" was the first object of its kind to be discovered (thanks to the excessive infrared radiation it emits, as seen in the image above).

	R.A.	DEC.	TYPE	MAGNITUDE	DISTANCE
β Pic	05h 47m	-51°04"	Planetary system	3.9	63 l.y.
δ Pic	06h 10m	-54°58"	Binary star	4.7–4.9 (var)	1,650 l.y.
β Dor	05h 34m	-62°29"	Variable star	3.5–4.1	1,040 l.y.
LMC	05h 24m	-69°45"	Irregular galaxy	0.1	179,000 l.y.
NGC 2070	05h 39m	-69°06"	Diffuse nebula	8.0	180,000 l.y.

IRREGULAR SATELLITE

The Large Magellanic Cloud (LMC) is the largest of the Milky Way's satellite galaxies, following a 1.5 billion-year orbit shared with the Small Magellanic Cloud (see page 216). With a diameter of 20,000 light years, it resembles an isolated patch of our own galaxy, 20 times the width of the full Moon, on the border of Dorado and Mensa. The LMC is easily spotted with the naked eye, and binoculars reveal a broad bar of stars that is its major feature—although usually classed as irregular, the LMC shows some structure and is sometimes called a "one-armed spiral." Telescopes of any size are ideal for exploring its nebulae and star clusters.

R.A. 05h 24m, Dec. -69°45"
Magnitude 0.1
Distance 179,000 light years

STARRY VISTA
LH 95

Although dwarfed by the nearby Tarantula Nebula, this star-forming region of the LMC has provided a wealth of information about the way stars form inside this gas- and dust-rich galaxy. Until recently, the nebula was known only from bright and hot blue-white stars, with up to three times the mass of the Sun. In 2006, however, a study using the Hubble Space Telescope revealed more than 2,500 infant stars that have not yet settled down onto the "main sequence" of stellar evolution (the yellow and orange stars in the image above). These include red dwarfs with as little as one-third the mass of the Sun.

R.A. 05h 37m, Dec. -66°22"
Magnitude 11.1
Distance 180,000 light years

TARANTULA NEBULA
NGC 2070

The Tarantula Nebula is one of the largest star-forming regions in our Local Group of galaxies. Binoculars or a small telescope will show it as a collection of long gaseous tendrils resembling the legs of a spider. Overall, the complex is around 1,000 light years across—if it was transplanted to the current location of the Orion Nebula M42, it would cover some 30 degrees of the sky and be bright enough to cast shadows. The nebula is thought to owe its size and intensity to its position on the leading edge of the LMC, where it suffers from compression as the galaxy moves around its orbit.

R.A. 05h 39m, Dec. -69°06"
Magnitude 8.0
Distance 180,000 light years

▶ SUPERSTARS
R136

Small telescopes are enough to spot the star cluster at the heart of the Tarantula Nebula, but they cannot hint at its awesome scale. This dense ball of heavyweight blue stars is just 1–2 million years old and pumps out intense ultraviolet radiation that excites gas throughout the enormous nebula. At its heart sits a tight knot of stars called R136a, recently separated into individual components that include R136a1, the most massive star known. This stellar monster has 265 times the mass of the Sun and 10 million times its luminosity— it is about as big as a star can get without tearing itself apart.

R.A. 05h 39m, Dec. -69°06"
Magnitude 9.5
Distance 180,000 light years

MENSA AND VOLANS

TABLE MOUNTAIN AND THE FLYING FISH

These two constellations of the far south are lacking in both bright stars and noticeable patterns, but are relatively easy to locate since Volans lies directly to the south of bright Carina, and Mensa lies between the Large Magellanic Cloud and the south celestial pole.

Volans was one of the new southern constellations invented by the Dutch navigators Keyser and de Houtman in the 1590s, while Mensa was added to the sky by Nicolas Louis de Lacaille in the 1750s, inspired by the daily sight of Table Mountain with its top shrouded in cloud.

Alpha Mensae is an unimpressive yellow dwarf with 80 percent of the Sun's luminosity, only visible to the naked eye, at magnitude 5.1, because it lies just 33 light years away. Volans at least has two attractive binaries to offer, both divisible with a small telescope. Epsilon Volantis is a blue-white star of magnitude 4.4 with a yellow companion of magnitude 8.1, while Gamma Volantis pairs an orange giant of magnitude 3.8 with a white star of magnitude 5.7.

THE MEATHOOK GALAXY At magnitude 11.2, this beautiful and unusual spiral is only visible through larger amateur telescopes, but it is nevertheless a highlight of far southern skies. Officially catalogued as NGC 2442, the galaxy's remarkable S-shape is formed by two asymmetric spiral arms, one of which is unusually compressed, while the other is elongated. The distortion is thought to be the result of a recent close encounter with another nearby galaxy, some 50 million light years from Earth.

	R.A.	DEC.	TYPE	MAGNITUDE	DISTANCE
γ **Vol**	07h 09m	-70°30"	Binary star	3.8/5.7	147 l.y.
ε **Vol**	08h 08m	-68°37"	Binary star	4.4/8.1	640 l.y.
NGC 2442	07h 36m	-69°32"	Spiral galaxy	11.2	50 million l.y.

CHAMAELEON AND APUS

THE CHAMAELEON AND THE BIRD OF PARADISE

These two far southern constellations are faint and shapeless, but easily located between the south celestial pole and the brighter stars of Carina, Musca, and Triangulum Australe.

Both these constellations are inventions of the Dutch navigators Keyser and de Houtman, who based many of their new star patterns on creatures they encountered on their trading mission to the East Indies. They were introduced to European astronomers through a celestial globe produced by Petrus Plancius in 1598.

While this area of the sky is devoid of deep-sky objects suitable for amateurs, the two constellations offer a number of double stars. Delta Chamaeleontis is a line-of-sight alignment consisting of stars at magnitudes 4.4 and 5.5, 355 and 365 light years from Earth respectively. Delta Apodis is a pairing of red and orange giants at magnitudes 4.7 and 5.3—both are around 700 light years away, and they may be a true binary. Epsilon Chamaeleontis is a binary pair of twin white stars, magnitudes 5.4 and 6.0, that can only be split with a medium-sized telescope.

CHAMAELEON COMPLEX The far southern skies might appear to be devoid of obvious star-forming nebulae, but in fact a broad swathe of dark clouds fills the constellation of Chamaeleon. This star-forming region some 500 light years from Earth remains inconspicuous because it is giving birth to lightweight, Sunlike stars, which do not excite the surrounding nebulosity enough to make it glow.

	R.A.	DEC.	TYPE	MAGNITUDE	DISTANCE
δ Cha	10h 45m	-80°30"	Visual double	4.4, 5.5	355, 365 l.y.
ε Cha	12h 00m	-78°13"	Binary star	5.4/6.0	365 l.y.
δ Apo	16h 20m	-78°41"	Visual double	4.7, 5.3	700 l.y.
θ Apo	14h 05m	-76°48"	Variable star	4.8–6.1	330 l.y.
NGC 3195	10h 10m	-80°25"	Planetary nebula	11.5	5,500 l.y.

HYDRUS
AND OCTANS
THE LITTLE WATER SNAKE
AND THE OCTANT

The region around the south celestial pole is sadly bereft of bright stars, marked by the obscure constellation of the Octant. In terms of objects of interest, the nearby Little Water Snake fares little better.

Hydrus forms a faint zigzag in the sky south of the bright star Achernar in Eridanus. Invented by the Dutch navigators Keyser and de Houtman, it was popularized by Petrus Plancius around 1600. Octans, meanwhile, is one of Lacaille's 18th-century inventions from the 1750s, named after a now-obsolete navigational instrument.

   Beta Hydri is one of the brightest stars in our solar neighborhood, 24 light years from Earth. It has 10 percent more mass than the Sun, but is somewhat older. Having exhausted its primary fuel supply, it has begun to swell and brighten in an effort to keep shining—a process that will eventually transform it into a brilliant red giant. Lambda Octantis is an attractive double with yellow and white components of magnitudes 5.4 and 7.7. It lies 435 light years from Earth and is easily separated in a small telescope.

POLAR STARS A stunning long-exposure view from the Australian coast shows the south celestial pole hanging over the Southern Ocean, with the horizon lit by a beautiful display of the Aurora Australis or Southern Lights. The sky's southern pivot is sadly lacking in a bright star equivalent to the northern Polaris—the closest naked-eye star to the pole itself is the obscure Sigma Octantis, but at magnitude 5.4 it is only visible under dark skies.

	R.A.	DEC.	TYPE	MAGNITUDE	DISTANCE
π **Hyi**	02h 14m	-68°20"	Visual double	5.6, 5.7	740, 470 l.y.
α **Oct**	21h 05m	-77°01"	Variable star	5.1 (var)	148 l.y.
λ **Oct**	21h 51m	-82°43"	Binary star	5.4/7.7	435 l.y.

THE WANDERING PLANETS

THE WORLDS OF OUR SOLAR SYSTEM ARE UNIQUE
AS THE ONLY CELESTIAL OBJECTS THAT CHANGE
THEIR POSITIONS OVER RELATIVELY SHORT
TIMESCALES, DANCING AMONG THE MORE DISTANT
STARS OF THE FIXED CONSTELLATIONS. THIS
CHAPTER LOOKS AT THESE BODIES IN DETAIL,
RANGING FROM OUR PLANET'S OWN MOON,
THROUGH THE EIGHT MAJOR PLANETS, TO MINOR
WORLDS SUCH AS ASTEROIDS AND COMETS. IT
INCLUDES ADVICE ON OBSERVING HIGHLIGHTS,
AND CHARTS FOR LOCATING KEY OBJECTS.

OUR SOLAR SYSTEM

EMPIRE OF THE SUN

Earth is just the third of eight major planets in orbit around the Sun, accompanied by countless smaller worlds including satellites, asteroids, and comets. The most prominent of these objects orbit in the plane of the solar system and are usually visible against the constellations of the zodiac.

The Sun is the dominant object in our skies—our closest star at a distance of 93 million miles (149.6 million kilometers). Scattered in Earth's atmosphere, its light turns the sky blue and drowns out the light of fainter objects. The visibility of the planets therefore depends on their angular separation from the Sun in our skies, and so there is a distinct difference in behavior between the "inferior" planets (Mercury and Venus) whose orbits lie inside Earth's, and the "superior" planet (all the rest) whose orbits lie beyond Earth. Inferior planets move back and forth from one side of the Sun to the other, while superior planets move all the way around the sky, tracking generally eastward against the zodiac, from the pre-dawn sky, through "opposition" when they are visible all night, before approaching the Sun again in the evening sky

MAP OF THE SOLAR SYSTEM
The planets fall into two distinct groups—the closely spaced small rocky planets of the inner Solar System, and the more widely spaced giant worlds of the outer Solar System. They all follow elliptical orbits that are more or less circular, and lie close to the flat plane known as the ecliptic, defined by Earth's motion around the Sun. As a result, they frequently come close to each other in Earth's skies, in events known as planetary conjunctions. The pages that follow look at the planets and other objects in order of their distance from the Sun.

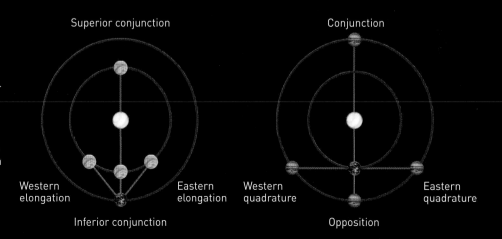

► PLANETARY ALIGNMENTS

For an Earth-based observer, planets move
in different ways depending on whether their
orbits are inferior (closer to the Sun) or superior
(further away from the Sun). Inferior planets
move between conjunctions on the near and far
sides of the Sun, passing through elongations
when they appear to the east or west of the Sun.
Superior planets have a single conjunction when
they lie on the far side of the Sun from Earth,
and an opposition when they are closest to
Earth and opposite the Sun in the sky.

Superior conjunction

Conjunction

Western
elongation

Eastern
elongation

Western
quadrature

Eastern
quadrature

Inferior conjunction

Opposition

THE MOON
EARTH'S SATELLITE

The Moon is the dominant object in Earth's skies after the Sun and has a huge influence on our planet. A constantly changing, slowly moving lantern in the sky, it is a rewarding target for amateur astronomers.

With a diameter of 2,158 miles (3,474 kilometers), Earth's Moon is the fifth-largest satellite in the Solar System, and by far the largest compared to its parent planet. Orbiting at an average distance of 239,000 miles (384,000 kilometers), it has an angular diameter of around half a degree in the sky, making it an ideal target for both binoculars and telescopes of any size. Its gravitational pull interacts with that of the Sun to produce rising and falling tides in Earth's oceans, and the full Moon at its brightest can drown out all but the brightest stars in the sky.

Even for naked-eye observers, the Moon's surface displays clear variations—patterns that various cultures have identified as a man, woman, or even a rabbit in the Moon are formed from dark areas known as seas—volcanic plains formed in low-lying

LUNAR MOVEMENTS

Tidal forces between Earth and the Moon have long since slowed our satellite's rotation so that it spins on its axis in precisely the same time it takes to complete an orbit. As a result, it keeps one hemisphere permanently turned toward Earth, and the other face forever hidden from view. As the Moon passes through its complete cycle of phases

surface are illuminated from different angles. The best telescopic views of the lunar landscape are generally obtained around local sunrise and sunset, when even low-relief features cast long shadows. As seen from Earth, these regions correspond to the "terminator" line between the Moon's dark and light sides.

FACE OF THE MOON

The Earth-facing near side of the Moon is a mix of bright, heavily cratered highlands and darker, comparatively smooth lava plains called "seas." Bright lines or "rays" crossing the surface preserve jets of material thrown across the lunar surface during the formation of relatively recent craters. With no atmosphere, water, or geological activity, cratering has been the only force shaping the lunar surface for around 3 billion years.

Seas:
1. Mare Cognitum
2. Mare Crisium
3. Mare Fertilitatis
4. Mare Frigoris
5. Mare Humorum
6. Mare Imbrium
7. Mare Nectaris
8. Mare Serenitatis
9. Mare Tranquilitatis
10. Mare Nubium
11. Mare Orientale
12. Oceanus Procellarum

Major craters:
13. Aristarchus
14. Plato
15. Copernicus
16. Ptolemaeus
17. Tycho

LUNAR CYCLES
PHASES AND ECLIPSES

The Moon's movement against the stars is the most noticeable change in the heavens from night to night. Over 29.5 days, our satellite goes through a cycle of phases, from new Moon to full Moon and back, as it returns to the same position, relative to the Sun, in the sky. But this is simply the most obvious among several patterns of motion.

The motions of the Moon are complicated by the fact that our satellite's orbit is tilted at 5.1 degrees to the plane of the ecliptic, and the orientation of this tilt slowly changes over time. So while the full "Harvest Moon" famously hangs close to the horizon on late summer nights, its precise altitude can vary by 5.1 degrees on either side of the ecliptic. Prehistoric astronomers realized the cycle between extreme "standstill" points repeats every 18.6 years. However, the Earth, Moon, and Sun return to their same relative positions in the sky in a slightly shorter period of 18 years and 11 days, an interval known as the Saros. After each Saros cycle, Sun and Moon are back at the same relative locations in the sky— and as a result the Saros is crucial to predicting the celestial alignments that produce eclipses.

SOLAR ECLIPSE

Solar eclipses are the most spectacular events in nature—a result of the Moon appearing almost exactly the same size as the Sun in Earth's skies. On rare occasions (no more than once or twice a year), the Moon passes across the face of the Sun as seen from Earth. Where it blocks the Sun completely it casts a small, deep shadow (the "umbra") that sweeps across Earth's surface at high speed. Beneath the umbra, the Sun's bright disk may be hidden for several minutes, revealing its tenuous atmosphere or corona. Elsewhere on Earth, observers witness only a partial eclipse. When the situation is reversed, the Moon can pass through Earth's own shadow, turning red or copper-colored in a lunar eclipse as direct sunlight is blocked from its surface.

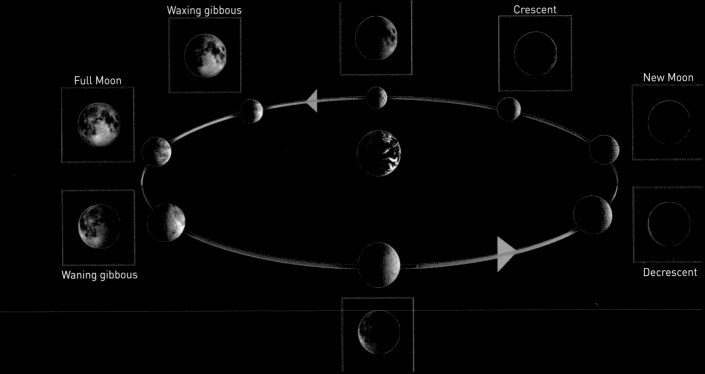

and in consequence the fastest-moving, completing an orbit in just 88 days. Never appearing far from the Sun, it is a fleeting visitor to the skies around dawn or dusk—often lost in haze despite its relative brightness.

Mercury is also the planet with the most elliptical orbit—its distance from the Sun ranges from 28.5–43.5 million miles (46–70 million kilometers). As a result, there can be considerable differences in its separation from the Sun from one elongation to the next. Even assiduous searchers can usually only spot it a couple of times a year, when it is visible against relatively dark skies and reaches a maximum brightness of around magnitude -2.7.

Just as with the Moon and Venus, we see different amounts of Mercury's sunlit surface as we view it from different angles, so the planet goes through distinct phases. However, the planet's tiny size means that these phases are only visible through a

TINY PLANET

Mercury is the smallest planet, with a diameter roughly 50 percent larger than Earth's Moon. Spaceprobe photographs reveal a similarly airless and cratered surface, but Mercury has signs of a more interesting past, with traces of volcanic

largest crater in the Solar System, with a diameter of 950 miles (1,550 km). Mercury's rotation is locked so that it spins three times on its axis for every two orbits around the Sun—most parts of the planet experience as a result

LOCATING MERCURY

The chart below tracks the motion of Mercury through the zodiac constellations over the next decade, showing its location at favourable elongations. A letter "E" indicates an eastern (evening) apparition, while "W" indicates a western (morning) elongation.

PHASES

Seen from Earth, Mercury passes through a Moon-like cycle of phases over the period of 115.88 days between inferior conjunctions with Earth. Around greatest elongations, its disk is roughly 50 percent illuminated—crescent and gibbous phases are increasingly hard to observe as Mercury always lies closer to the Sun at these times.

PLANETARY TRANSITS

As seen from Earth, each of the inferior planets can occasionally pass across the face of the Sun in an event called a transit. The planet forms a small circular dot on the Sun, moving slowly from east to west, and best seen by safely projecting the Sun's image onto a screen. Never look at the Sun directly through any optical instrument.

VENUS

MORNING AND EVENING STAR

Brilliant Venus is the brightest object in Earth's skies after the Sun and Moon—usually the closest planet to Earth, and our near-twin in terms of size. Its searing, toxic atmosphere reflects most of the sunlight that falls upon it, turning it into a blazing beacon in the skies after sunset or before sunrise.

Venus orbits the Sun in 224.7 days at an average distance of 67 million miles (108 mllion kilometers). As a result, its distance from Earth ranges between a mere 26 million miles (42 million kilometers) at inferior conjunction and 160 million miles (258 million kilometers) at superior conjunction.

As the two planets move around their orbits and we see more or less of Venus's sunlit side, it goes through a cycle of phases and varies in brightness between magnitudes -4.9 and -3.0. Although its orbit is inferior to Earth, Venus can still get to almost 48 degrees from the Sun, rising well before it or setting well after it against dark skies. When brightest and closest to Earth, binoculars clearly show the planet's large crescent, but a telescope is needed to view its gibbous phase on the far side of its orbit from Earth.

EVIL TWIN

Venus's diameter of 7,518 miles (12,104 km) is just 5 percent smaller than that of Earth, but the similarities more or less end there. Venus is cloaked in a toxic atmosphere that reaches a crushing 100 times Earth atmospheric pressure at the surface. The carbon dioxide that dominates this atmosphere produces a runaway greenhouse effect that ensures an average surface temperature of 860°F (460°C). The planet's surface, meanwhile, is dominated by volcanoes and lava fields, which re-covered most of the landscape in a wave or eruptions around 600 million years ago, and may still be active today. Another strange feature is Venus's slow rotation—the planet spins on its axis in 243 days (longer than a Venusian year), effectively spinning backward.

LOCATING VENUS

The chart below tracks the motion of Venus through the zodiac constellations over the next decade at three-monthly intervals. Gaps in the sequence indicate when Venus is too close to the Sun to be seen. A letter "E" indicates an eastern (evening) elongation, while "W" indicates a western (morning) elongation.

AQUARIUS · CAPRICORNUS · SAGITTARIUS · OPHIUCHUS · SCORPIUS · LIBRA · VIRGO · LEO · CANCER · GEMINI · TAURUS · ARIES · PISCES

3/2024 (W)
3/2016 (W)

3/2019 (W)
3/2022 (W) 12/2016 (E)
3/2014 (W)

12/2013 (E)
12/2021 (E)
12/2019 (E)

12/2022 (E)

12/2020 (W)

12/2015 (W) 12/2023 (W)
12/2018 (W)
9/2021 (E) 9/2013 (E)
9/2018 (E)

9/2024 (E)
9/2016 (E)

9/2017 (W)

9/2023 (W) 9/2015 (W)
9/2020 (W)
6/2023 (E) 6/2015 (E)
6/2018 (E)

6/2021 (E)

6/2020 (W)
6/2019 (W)

6/2022 (W)
6/2014 (W)

3/2020 (E)
6/2017 (E)

3/2015 (E)
3/2023 (E)

3/2018 E)

EVENING STAR

Although ancient astronomers were fully aware that Venus was the same object during both eastern and western elongations, they recognized its two separate aspects as Lucifer and Hesper, the morning and evening stars (the name Lucifer translates as "bringer of light"). These days, the planet's brilliant appearance frequently prompts UFO reports.

VENUS EXPOSED

Since the dawn of the Space Age, astronomers have been able to see through Venus's dense atmosphere using radar signals, allowing them to produce the first maps of the planet. The speed with which different parts of the surface reflect radio waves back to an orbiting spacecraft indicates their relative height, while the intensity of the returning signal can reveal other geological properties.

MARS

THE RED PLANET

Unmistakable thanks to its distinctive color, Mars is actually the most Earth-like world in our solar system, with similarities that get more obvious with each successive spaceprobe mission.

Orbiting directly beyond Earth, Mars has the most complex path through Earth's skies of any planet. The Red Planet completes a circuit of the Sun every 687 Earth days, and has a distinctly elliptical orbit that ranges between 34 million and 61 million miles (207 million and 249 million kilometers) from the Sun. As a result, even at opposition (closest approach to the Earth), Mars ranges between 34 million and 61 million miles (55 –100 million kilometers) away. What's more, Mars makes large retrograde or backward loops in the sky as Earth overtakes it in its orbit around each opposition.

Mars ranges from magnitude -3.0 at its brightest to +1.6 at minimum, and varies hugely in appearance. At closest approach binoculars will clearly show it as a small disk, while small telescopes will reveal dark and light markings across its surface—hints of the planet's icy polar caps and dusty plains.

COLD DESERT

With a diameter of 4,219 miles (6,792 km), Mars is a little more than half the size of Earth. Its cold, dry atmosphere is dominated by carbon dioxide, but exerts just a thousandth of the pressure of Earth's air at the surface. As a result, Mars today is a cold and arid planet, red in color thanks to the iron oxide (rust) prevalent in its fine soils. There is a distinct dichotomy between lowland deserts in the northern hemisphere and cratered highlands in the south, but the planet's most impressive features are towering volcanic peaks, including Olympus Mons, the tallest mountain in the Solar System at 19 miles (27 km) high. Equally awe-inspiring is the Valles Marineris canyon system, which is up to 6 miles (10 km) deep in places and more than 2,500 miles (4,000 km) long.

LOCATING MARS

The chart below tracks the motion of Mars through the zodiac constellations over the next decade at three-monthly intervals. Gaps in the sequence indicate when Mars is too close to the Sun to be seen.

AQUARIUS · CAPRICORNUS · SAGITTARIUS · OPHIUCHUS · SCORPIUS · LIBRA · VIRGO · LEO · CANCER · GEMINI · TAURUS · ARIES · PISCES

6/2020 · 12/2018 · 12/2016 · 3/2024 · 12/2014 · 6/2018 · 3/2022 · 9/2018 · 3/2020 · 3/2018 · 9/2016 · 12/2023 · 3/2016 · 12/2021 · 9/2014 · 6/2016 · 12/2019 · 12/2017 · 12/2013 · 9/2021 · 3/2014 · 12/2015 · 6/2014 · 9/2023 · 9/2019 · 9/2017 · 9/2015 · 6/2023 · 9/2013 · 12/2024 · 6/2021 · 6/2019 · 6/2017 · 9/2024 · 3/2023 · 6/2015 · 12/2022 · 9/2022 · 3/2021 · 3/2019 · 6/2024 · 3/2017 · 9/2020 · 12/2020 · 6/2022 · 3/2015

MARTIAN MARKINGS

The Red Planet's most prominent features, visible through small telescopes around opposition, are its bright polar caps—a mix of water ice and frozen carbon dioxide. Darker markings, which are best viewed using colored filters, vary in size and intensity from month to month as the Martian weather shifts fine dust across its surface.

RETROGRADE MOTION

Mars, like all the superior planets, drifts east against the stars through each orbit. But as the faster-moving Earth overtakes its outer neighbor (roughly every 26 months, around the time of opposition) our changing point of view causes Mars to slow down its motion, and eventually reverse, describing a large westward or "retrograde" loop in the sky. All superior planets display this kind of retrograde motion, but the loops followed by Mars are by far the largest.

THE ASTEROIDS

DEBRIS OF THE SOLAR SYSTEM

Between the orbits of Mars and Jupiter lies a swarm of billions of small rocky bodies left over from the formation of the Solar System. While the majority of these tiny worlds are too faint to see with amateur telescopes, a handful can be easily tracked and even spotted with the naked eye.

Italian astronomer Giuseppe Piazzi spotted the first asteroid, Ceres, in 1801 while looking for a missing planet in the perceived gap between Mars and Jupiter. Within a few years, three more—Pallas, Juno, and Vesta—were discovered, but it was several decades before the floodgates opened to the discovery of many more. The main Asteroid Belt is largely confined in the plane of the ecliptic, so asteroids are usually found in or around the zodiac constellations.

Ceres and Vesta are the two best asteroids for amateurs, and among the most interesting. Varying in magnitude between 6.6 and 9.3, Ceres can usually be spotted with binoculars. Vesta is smaller, but orbits closer to the Sun and has a brighter surface, so it varies between magnitudes 5.1 (visible to the naked eye) and 8.5.

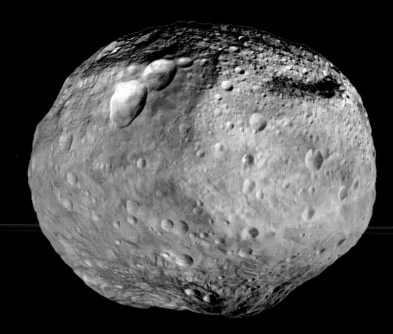

VESTA

With a maximum diameter of 355 miles (572 km), Vesta is one of the largest asteroids and thanks to NASA's Dawn spaceprobe, one of the most intensively studied. However, it seems to be unusual among the asteroids, since it has a bright surface covered in minerals suggesting past volcanic activity (smaller asteroids, in contrast, have led cold inactive lives and many are thought to be more or less unaltered samples of the Solar System's raw materials). A huge crater called Rheasilvia, some 314 miles (505 km) in diameter, dominates the asteroid's southern hemisphere. Formed in a massive impact, it allows scientists to peer into the depths of Vesta's interior.

LOCATING CERES AND VESTA

The chart below tracks the motions of the prominent asteroids Ceres (purple) and Vesta (gray) through the zodiac constellations over the next decade at six-monthly intervals. Gaps in the sequence indicate when they are too close to the Sun to be seen.

AQUARIUS · CAPRICORNUS · SAGITTARIUS · OPHIUCHUS · SCORPIUS · LIBRA · VIRGO · LEO · CANCER · GEMINI · TAURUS · ARIES · PISCES

6/2020 · 12/2020 · 12/2022 · 6/2022 · 12/2015 · 6/2015 · 12/2018 · 12/2024 · 6/2024 · 12/2015 · 12/2019 · 12/2014 · 6/2018 · 12/2014 · 12/2021 · 12/2023 · 6/2019 · 12/2017 · 12/2018 · 6/2014 · 12/2024 · 6/2014 · 12/2013 · 12/2013 · 12/2022 · 6/2023 · 6/2021 · 12/2020 · 6/2018 · 12/2017 · 6/2017 · 12/2016 · 6/2024 · 6/2022 · 6/2020 · 12/2023 · 6/2017 · 6/2016 · 12/2021 · 6/2023 · 6/2021 · 12/2019 · 6/2019 · 6/2016 · 12/2016 · 6/2015

CERES

The largest asteroid of all, Ceres is technically classed as a "dwarf planet" on account of its spherical shape. With a diameter of 590 miles (950 km), Ceres contains one-third of the mass of the entire asteroid belt. Images such as this Hubble Space Telescope view suggest that it has a complex surface, with hints of an atmosphere and patches of ice.

NEAR EARTH OBJECTS

Not all asteroids are confined to the main Asteroid Belt—considerable numbers follow elliptical orbtis that come closer to the Sun, crossing the orbits of Mars and sometimes Earth. These "Near Earth Objects," such as 433 Eros, shown here, are among the easiest of all asteroids to study, and can reach naked-eye or binocular visibility when at their closest to Earth.

JUPITER
KING OF THE PLANETS

Jupiter is the nearest and largest of our solar system's giant planets—so large that, despite remaining at least 380 million miles (600 million kilometers) from Earth at all times, it is still the brightest object in the sky after the Sun, Moon, and Venus.

This giant world takes 12 years to orbit the Sun, and as a result moves relatively slowly against the more distant stars, taking a year or so to travel through each of the major zodiac constellations. Like all the superior planets, it displays retrograde loops in its path across the sky, but these are both smaller and simpler than the ones displayed by Mars.

Binoculars are enough to reveal a small but distinct disk, with the star-like points of the planet's four large "Galilean" moons lined up on either side. The smallest telescope, meanwhile, will show the planet's bulging shape, and reveal the most prominent of its contrasting cloud bands. The visibility of the famous Great Red Spot, however, varies considerably from year to year. Larger instruments will show copious detail among the clouds, making Jupiter one of the most rewarding objects in the sky.

GAS GIANT

Jupiter has a very different structure from the rocky planets of the inner Solar System—it is the first of the gas giants. As this name suggests, it is a world dominated by its atmosphere. This enormous envelope of lightweight gas (predominantly hydrogen) compresses to form a global ocean below a certain depth and surrounds a solid core roughly the size of Earth. Jupiter's equatorial diameter is 11.2 times that of our own planet, but it rotates in just ten hours, causing it to bulge at the equator compared to the poles. Weather systems in the upper atmosphere are colored by a variety of chemicals, creating a pattern of alternating cream-colored "zones" and reddish "belts" parallel to the equator.

AQUA...   CAPRI...   SAGITTARIUS   SCORPI...   LIBRA   VIRGO...

12/2022   6/2021   12/2021   12/2020   6/2020   12/2019   6/2019   12/2018   12/2017   6/2018   12/2016   6/2017   12/2015   6/2016   12/2014   6/2015   6/2014   12/2013   12/2024   6/2024   6/2023   12/2023   6/2022

GREAT RED SPOT

Jupiter's most famous feature is an enormous anticyclonic storm, large enough to swallow up the Earth, rotating in the southern hemisphere. The Great Red Spot is colored by chemicals drawn up from deep inside the atmosphere, and varies in intensity over time—at times it can merge into its surroundings, leaving just a "hollow" marking its position.

SATELLITE FAMILY

Four huge worlds stand out among Jupiter's family of 67 satellites. These "Galilean" moons—Io, Europa, Ganymede, and Callisto—orbit with periods ranging from 1.8 to 16.7 days, in an ever-changing dance. Small telescopes can show the frequent events in which the moons disappear behind Jupiter or move in front of it, casting their shadows onto the clouds.

Saturn, the outermost of the planets known to ancient stargazers, makes its stately progress through the zodiac constellations in 29.5 years. Its chief attraction for amateur astronomers lies in its intricate and beautiful system of rings.

Orbiting an average of 870 million miles (1.4 billion kilometers) from the Sun, Saturn varies in brightness between magnitudes -0.24 and +1.5 depending on its position relative to Earth. Its impressive ring system lies directly above its equator, but because Saturn is tilted on its axis by 27 degrees, we see it from different angles at different points in its orbit.

Most of the time, binoculars will show that Saturn is not spherical, while telescopes of various grades will show increasing amounts of detail in the rings. At their most open to Earth, they extend from the planet like a pair of handles, while at other times they form a thinner oval aligned with the equator. Twice in each Saturnian year, the rings lie edge-on to Earth and are invisible to all but the most powerful telescopes. Larger instruments may also

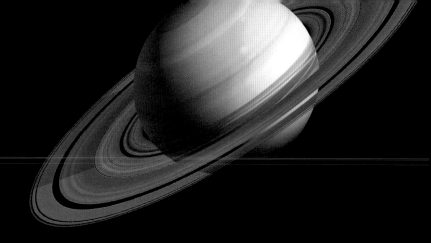

RINGED WONDER

Saturn is a gas giant like Jupiter, but with some key differences. It is the least dense planet in the Solar System (less dense on average than water), and bulges at the equator even more obviously than Jupiter. A creamy-white haze of ammonia crystals formed in the cold upper atmosphere hides most of its weather systems from view, although

from Earth. The rings consist of countless particles, ranging in size from house-sized boulders to specks of dust and dominated by highly reflective water ice. Collisions and jostling among the particles, coupled with the gravitational influence of Saturn's many satellites, orders them into countless ringlets with relatively empty regions between them

LOCATING SATURN

The chart below tracks the motion of Saturn through the zodiac constellations over the next decade at six-monthly intervals. Gaps in the sequence indicate when Saturn is too close to the Sun to be seen.

AQUARIUS   CAPRICORNUS   SAGITTARIUS   OPHIUCHUS   SCORPIUS   LIBRA   VIRGO   LEO   CANCER   GEMINI   TAURUS   ARIES   PISCES

6/2024 · 12/2024 · 6/2023 · 12/2023 · 6/2022 · 12/2022 · 6/2021 · 12/2021 · 6/2020 · 12/2020 · 12/2019 · 6/2019 · 12/2018 · 6/2018 · 12/2017 · 6/2017 · 12/2016 · 6/2016 · 12/2015 · 6/2015 · 12/2014 · 12/2013 · 6/2013

RING CYCLE

A series of images from the Hubble Space Telescope tracks the changing aspect of Saturn's rings from almost edge-on to Earth in late 1996, to wide-open in late 2000. During this period, Saturn's southern hemisphere moved from spring into summer as it received increasing amounts of sunlight. The northern hemisphere, meanwhile, moved from fall into winter.

TITAN

Saturn has a huge family of at least 62 moons, several of which are complex worlds in their own right. The system is dominated by Titan, the second largest satellite in the Solar System with a diameter of 3,200 miles (5,152 km) and a thick, methane-rich atmosphere. With a magnitude between 8.2 and 9.0, Titan can usually be seen close to Saturn through binoculars or a small telescope.

URANUS
WORLD ON ONE SIDE

Uranus was the first planet to be discovered in the telescopic age, by German-born British astronomer William Herschel in 1781. Yet, in fact, it lies just on the limits of naked-eye visiblity—if you know where to look.

Slow-moving Uranus takes a sedate 84 years to orbit the Sun at an average distance of 1.8 billion miles (2.9 billion kilometers). From Earth, its brightness varies between magnitudes 5.9 and 5.32, so it should be visible to the naked eye in dark skies, with the aid of a finder chart such as the one shown opposite. Through binoculars it appears as a turquoise star-like point.

At less than half the size of Saturn and roughly twice its distance, Uranus appears as a small, featureless disk through most amateur telescopes. Larger instruments may reveal the planet's brightest moons, Titania and Oberon, at around magnitude 14.0 Careful tracking of the moons as they orbit their parent planet reveals Uranus's most unusual feature—the entire system is tipped on one side so that its axis of rotation lies close to the plane of the ecliptic.

ICE GIANT

While Jupiter and Saturn are gas giants, astronomers often refer to Uranus and Neptune as "ice giants." Smaller and denser than their inner neighbors, the inner layers of these planets are composed of "slushy" mixes of materials such as water and ammonia—compounds with low melting points that are chemically known as ices.

The outer atmospheres, meanwhile, are dominated by lightweight hydrogen and helium, with small amounts of methane that absorbs red light and gives both worlds their distinctive color. Uranus's strange orientation may have been caused by a series of large impacts early in its history, or by the gravitational influence of the other planets during a series of ancient close encounters.

TAURUS

ARIES

ε

δ

β

γ₂

χ

η

PISCES

CETUS

ο

δ

ε

ν

μ

α

ξ

12/2024
12/2023
12/2022
12/2021
12/2020
12/2019
12/2018
12/2017
12/2016
12/2015
12/2014

LOCATING URANUS

This chart tracks the motion of Uranus through the constellations of Pisces and Aries over the next decade at three-monthly intervals. Gaps in the sequence indicate when Uranus is too close to the Sun to be seen.

CHANGING MOODS

Uranus's tilt gives it a unique climate, with most of the planet cycling between a sunless winter and a summer of endless daylight that each last for decades. Extreme temperature differences between the hemispheres suppress the planet's weather, so Uranus only develops obvious storms and cloud bands around the equinoxes, when sunlight is more evenly distributed.

RINGSIDE VIEW

Uranus has a system of rings that are very different from those around Saturn. Thirteen narrow ringlets of dark material, rich in methane ice, encircle the planet's equator. Discovered in 1977 when their passage in front of a distant star caused its light to flicker, they are only visible from Earth through the most powerful professional telescopes.

NEPTUNE
OUTERMOST GIANT

Distant Neptune is the outermost of
the major planets (tiny Pluto, which lies
beyond it, is now considered to be simply
the brightest among countless "ice dwarf"
worlds ringing the outer Solar System).
It was discovered in 1846 by French
mathematician Urbain Le Verrier,
who predicted its location from its
gravitational influence on Uranus.

Neptune takes 164.8 years to complete an orbit
of the Sun, at an average distance of 2.8 billion miles
(4.5 billion kilometers). It varies little in brightness,
between magnitudes 7.8 and 8.2, so is relatively easy
to locate with binoculars and unmistakable through
a small telescope. Despite this, even the largest
amateur telescopes will show no more than a small
disk, distinctly bluer in hue than the otherwise
similar Uranus.

Despite its distance, Neptune's major moon
Triton is actually easier to see than the large
moons of Uranus—it reflects more sunlight back to
Earth thanks to its considerable size. Nevertheless,
at magnitude 13.5, Triton can still only be seen
through medium-sized instruments.

STORMY WORLD

Neptune is slightly smaller than
Uranus, and has about 50 percent more
methane in its atmosphere. This means
that it absorbs considerably more of
the red component of sunlight striking
its upper layers, and therefore appears
a more intense shade of blue. With an
axial tilt similar to Earth, it spins once
every 16 hours, and displays surprisingly
complex and powerful weather systems.

Despite their superficial similarities
to Jupiter's Great Red Spot, however,
Neptune's dark atmospheric spots are
quite short-lived features. Three thin
rings encircle the planet, but they are
much less impressive than those of
either Saturn or even Uranus.

PEGASUS

19

γ   β

λ

κ   PISCES

η   ζ₁   π

α

γ

ο

27

φ   AQUARIUS   θ

33.30

CETUS   χ

ψ₂   ψ₁   λ

12/2024   12/2023   12/2022   12/2021   12/2020   12/2019   12/2018   12/2017   12/2016   12/2015   12/2014   12/2013

τ₂

δ

ι

CAPRICORN

LOCATING NEPTUNE
This chart tracks the motion of Neptune through the
constellations of Aquarius and Pisces over the next decade
at three-monthly intervals. Gaps in the sequence indicate
when the planet is too close to the Sun to be seen.

COMETS
FROZEN VISITORS

Wanderers in the Solar System, comets
are tiny bodies composed of rock and ice.
They become prominent only during their
fleeting visits to the inner Solar System,
when the Sun heats their surfaces and
causes the volatile chemicals within to
evaporate into space.

Comets typically follow long, highly elliptical orbits
that ensure they spend most of their lives in the cold
outer Solar System, and achieve high speeds as they
swing past Earth and around the Sun. Ultimately,
they originate in the Oort Cloud, an enormous halo
containing trillions of frozen comet "nuclei" formed
in the early days of the Solar System. Occasional

COMET TAILS

Gas and dust streaming away from a comet's nucleus may interact with radiation from the Sun and the solar wind of particles blowing out across the solar system, to form one or more tails. Typically, gas ionized by solar radiation forms a glowing, bluish "ion tail" that points straight away from the Sun regardless of the direction of the comet's motion. The "dust tail," meanwhile, shines by reflecting sunlight: while it, too, generally points away from the Sun, it tends to curve along the direction of the comet's orbit.

CENTRAL STRUCTURE

As a comet nucleus approaches the Sun, gas evaporating from its surface and escaping from its interior can develop into a huge but tenuous atmosphere, still trapped by the weak gravity of the nucleus. This vast gas envelope, known as the coma, may grow to be larger than the Sun—the spectacular example shown here developed around Comet Holmes in 2007.

METEOR SHOWERS

Fragments of ice and dust ejected during a comet's passage around the Sun continue to follow the same orbit, and in a few cases Earth's own passage round the Sun takes it straight through these clouds of particles. The result is a shower of shooting stars or meteors, all apparently coming from the same part of the sky (the "radiant") at the same time of year. See page 311 for a table of key meteor showers.

THE MONTHLY SKIES

WHILE OUR VIEW OF THE SKY CHANGES FROM
HOUR TO HOUR AND FROM NIGHT TO NIGHT, IT
REMAINS ESSENTIALLY IDENTICAL AT THE SAME
POINT IN EACH YEAR. THE CHARTS IN THIS
CHAPTER SHOW THE CHANGING ASPECT OF
EVENING SKIES FROM MONTH TO MONTH IN BOTH
NORTHERN AND SOUTHERN HEMISPHERES. EACH
MONTH IS REPRESENTED BY A PAIR OF MAPS
SHOWING THE VIEW LOOKING TOWARD THE
NORTHERN AND SOUTHERN HORIZONS FOR A
RANGE OF LATITUDES.

JANUARY

These charts show the northern hemisphere skies during January—they are correct for 11 p.m. local time (midnight if Daylight Saving Time is in effect) on January 1, 10 p.m. local time on January 15, and 9 p.m. local time on January 31.

LOOKING NORTH

20°N **40°N** 60°N

On January evenings Orion and his retinue of bright constellations dominate the skies looking south (right). The Milky Way runs from north to south across the sky, while Ursa Major and Cassiopeia lie on either side of the north celesial pole (left).

LOOKING SOUTH

FEBRUARY

These charts show the northern hemisphere skies during February—they are correct for 11 p.m. local time (midnight if Daylight Saving Time is in effect) on February 1, 10 p.m. local time on February 14, and 9 p.m. local time on February 28.

GEMINI

CANCER

LEO

AURIGA

LYNX

LEO MINOR

URSA MAJOR

TAURUS

PERSEUS

CAMELO-PARDALIS

URSA MINOR

CANES VENATICI

VIRGO

TRIANGULUM

CASSIOPEIA

CEPHEUS

COMA BERENICES

ARIES

ANDROMEDA

DRACO

BOOTES

LACERTA

CYGNUS

HERCULES

CORONA BOREALIS

PISCES

PEGASUS

LYRA

W

E

LOOKING NORTH

20°N **40°N** 60°N

As Orion moves westward, the "Winter Triangle" of Betelgeuse, Procyon, and Sirius becomes prominent in the southern sky (right), with the fainter constellations around Hydra to its east. Looking overhead, the bright zodiac constellations Gemini and Leo flank the fainter Cancer.

LOOKING SOUTH

E

W

MARCH

These charts show the northern hemisphere skies during March—they are correct for 11 p.m. local time (midnight if Daylight Saving Time is in effect) on March 1, 10 p.m. local time on March 15, and 9 p.m. local time on March 31.

CANCER

LEO

LEO MINOR

GEMINI

URSA MAJOR

COMA BERENICES

LYNX

CANES VENATICI

ORION

AURIGA

CAMELOPARDALIS

URSA MINOR

BOOTES

PERSEUS

TAURUS

CEPHEUS

DRACO

CORONA BOREALIS

CASSIOPEIA

SERPENS CAPUT

TRIANGULUM

ARIES

LACERTA

CYGNUS

LYRA

ANDROMEDA

HERCULES

W

E

LOOKING NORTH

As Orion sinks out of view, Leo dominates the spring sky looking south (right) with the familiar stars of Ursa Major above it and directly overhead. Hydra is around its highest point in the sky while the Milky Way is close to its lowest in the north (left).

URSA MAJOR

LYNX

CANES VENATICI

LEO MINOR

AURIGA

COMA BERENICES

CANCER

GEMINI

BOOTES

LEO

CANIS MINOR

TAURUS

SEXTANS

HYDRA

CRATER

MONOCEROS

ORION

VIRGO

CORVUS

PUPPIS

PYXIS

CANIS MAJOR

ANTLIA

LEPUS

E

W

LIBRA

CENTAURUS

VELA

ERIDANUS

COLUMBA

CARINA

PICTOR

LOOKING SOUTH

APRIL

These charts show the northern hemisphere skies during April—they are correct for 11 p.m. local time (midnight if Daylight Saving Time is in effect) on April 1, 10 p.m. local time on April 15, and 9 p.m. local time on April 30.

LEO

COMA BERENICES

LEO MINOR

CANES VENATICI

URSA MAJOR

BOÖTES

CANCER

SERPENS CAPUT

CORONA BOREALIS

LYNX

GEMINI

URSA MINOR

DRACO

AURIGA

CAMELOPARDALIS

CEPHEUS

LYRA

HERCULES

ORION

TAURUS

PERSEUS

CASSIOPEIA

LACERTA

CYGNUS

W

E

TRIANGULUM

ARIES

ANDROMEDA

VULPECULA

SAGITTA

LOOKING NORTH

20°N **40°N** 60°N

The brilliant stars Spica and Arcturus, in Virgo and Boötes respectively, are prominent in southern skies (right), and Ophiuchus rises from the east. To the north (left) Cassiopeia is at its lowest point in the sky, while Castor and Pollux in Gemini hang above the western horizon.

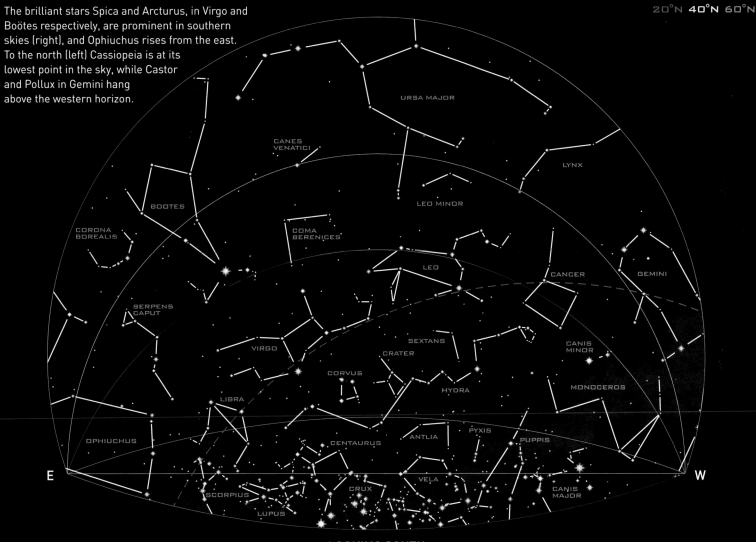

URSA MAJOR

CANES VENATICI

LYNX

BOOTES

LEO MINOR

CORONA BOREALIS

COMA BERENICES

LEO

CANCER

GEMINI

SERPENS CAPUT

CANIS MINOR

VIRGO

SEXTANS

CRATER

CORVUS

MONOCEROS

HYDRA

LIBRA

ANTLIA

PYXIS

PUPPIS

OPHIUCHUS

CENTAURUS

E

W

VELA

SCORPIUS

CRUX

CANIS MAJOR

LUPUS

LOOKING SOUTH

MAY

These charts show the northern hemisphere skies
during May—they are correct for 11 p.m. local time
(midnight if Daylight Saving Time is in effect)
on May 1, 10 p.m. local time on May 15,
and 9 p.m. local time on May 31.

LOOKING NORTH

20°N **40°N** 60°N

On early summer nights, Boötes and Hercules ride high in the southern skies (right), with galaxy-rich Coma Berenices and Virgo nearby. Ophiuchus and some of the richest Milky Way starfields are rising from the eastern horizon, while Leo sinks into the west.

LOOKING SOUTH

JUNE

These charts show the northern hemisphere skies during June—they are correct for 11 p.m. local time (midnight if Daylight Saving Time is in effect) on June 1, 10 p.m. local time on June 15, and 9 p.m. local time on June 31.

LOOKING NORTH

20°N **40°N** 60°N

By northern midsummer, Scorpius and Sagittarius are approaching their highest point above the southern horizon (right), bringing with them the richest regions of the Milky Way. To their northeast lies the rising northern "Summer Triangle" of Altair, Vega, and Deneb.

DRACO

CANES VENATICI

LYRA

CORONA
BOREALIS

BOOTES

CYGNUS

COMA
BERENICES.

LEO MINOR

VULPECULA

HERCULES

SAGITTA

SERPENS
CAPUT

LEO

DELPHINUS

OPHIUCHUS

VIRGO

AQUILA

SCUTUM

SERPENS
CAUDA

LIBRA

CORVUS

CRATER

SAGITTARIUS

LUPUS

SEXTANS

CAPRICORNUS

SCORPIUS

CORONA
AUSTRALIS

CENTAURUS

HYDRA

E

NORMA

W

TELESCOPIUM

ARA

LOOKING SOUTH

JULY

These charts show the northern hemisphere skies
during July—they are correct for 11 p.m. local time
(midnight if Daylight Saving Time is in effect)
on July 1, 10 p.m. local time on July 15,
and 9 p.m. local time on July 31.

LOOKING NORTH

20°N **40°N** 60°N

On northern July evenings, the Summer Triangle of Vega (in Lyra), Deneb (in Cygnus), and Altair (in Aquila) is at its highest in the southern sky (right). The Square of Pegasus rises from the east, while Ursa Major approaches its lowest point near the northern horizon (left) and the head of Draco is directly overhead.

DRACO

CYGNUS
LYRA
HERCULES
CORONA BOREALIS
BOOTES

VULPECULA
COMA BERENICES

SAGITTA
OPHIUCHUS
DELPHINUS
AQUILA
SERPENS CAPUT

PEGASUS
EQUULEUS
SCUTUM
SERPENS CAUDA
SAGITTARIUS
LIBRA
VIRGO

CAPRICORNUS
SCORPIUS

AQUARIUS
CORONA AUSTRALIS
LUPUS
CENTAURUS
HYDRA

E
MICROSCOPIUM
PISCIS AUSTRINUS
TELESCOPIUM
NORMA
CORVUS
W

INDUS
ARA
GRUS
PAVO

LOOKING SOUTH

271

AUGUST

These charts show the northern hemisphere skies during August—they are correct for 11 p.m. local time (midnight if Daylight Saving Time is in effect) on August 1, 10 p.m. local time on August 15, and 9 p.m. local time on August 31.

LOOKING NORTH

20°N **40°N** 60°N

In early August, the rich Perseid meteor shower radiates from Perseus as it hangs above the northeastern horizon (left). Meanwhile, the southern aspect (right) is dominated by the broad band of the Milky Way and the prominent dark dust lane known as the Cygnus Rift.

DRACO

LACERTA

CYGNUS

ANDROMEDA

LYRA

CORONA BOREALIS

VULPECULA

HERCULES

SAGITTA

PEGASUS

DELPHINUS

AQUILA

BOOTES

EQUULEUS

AQUARIUS

SCUTUM

SERPENS CAPUT

PISCES

CAPRICORNUS

OPHIUCHUS

SERPENS CAUDA

LIBRA

E

MICROSCOPIUM

SAGITTARIUS

SCORPIUS

VIRGO

W

CORONA AUSTRALIS

CETUS

PISCIS AUSTRINUS

GRUS

TELESCOPIUM

SCULPTOR

INDUS

PAVO

NORMA

LUPUS

ARA

LOOKING SOUTH

273

SEPTEMBER

These charts show the northern hemisphere skies
during September—they are correct for 11 p.m.
local time (midnight if Daylight Saving Time
is in effect) on September 1, 10 p.m. local
time on September 15, and 9 p.m.
local time on September 30.

LOOKING NORTH

20°N **40°N** 60°N

As the bright summer constellations linger in the west, the emptier skies around Aquarius and Pisces take up a dominant position in southern skies (right). The bright Square of Pegasus sits above them, while Cassiopeia and Perseus hang in the northeast (left).

LOOKING SOUTH

OCTOBER

These charts show the northern hemisphere skies during October—they are correct for 11 p.m. local time (midnight if Daylight Saving Time is in effect) on October 1, 10 p.m. local time on October 15, and 9 p.m. local time on October 31.

LOOKING NORTH

20°N **40°N 60°N**

Looking south, October skies seem relatively barren (right), though the Square of Pegasus is prominent and nearby Andromeda and Triangulum both contain spectacular galaxies. Turn in the other direction to admire the impressive arc of the Milky Way above the northern horizon (left).

LOOKING SOUTH

NOVEMBER

These charts show the northern hemisphere skies
during November—they are correct for 11 p.m.
local time (midnight if Daylight Saving Time is
in effect) on November 1, 10 p.m. local
time on November 15, and 9 p.m.
local time on November 30.

W

E

PISCES

ARIES

ANDROMEDA

TRIANGULUM

TAURUS

PEGASUS

PERSEUS

LACERTA

CASSIOPEIA

AURIGA

ORION

CEPHEUS

CAMELOPARDALIS

DELPHINUS

GEMINI

CYGNUS

LYNX

URSA
MINOR

VULPECULA

LYRA

CANCER

CANIS
MINOR

SAGITTA

DRACO

URSA
MAJOR

AQUILA

HERCULES

BOOTES

CANES
VENATICI

LEO
MINOR

LEO

CORONA
BOREALIS

40°N 60°N

Cetus, with its famous variable star Mira, dominates southern skies (right), while the rich constellations Taurus and Orion, rising from the east, herald the return of winter. The bright stars of Perseus and Cassiopeia lie directly overhead.

LOOKING SOUTH

DECEMBER

These charts show the northern hemisphere skies
during December—they are correct for 11 p.m.
local time (midnight if Daylight Saving Time
is in effect) on December 1, 10 p.m. local
time on December 15, and 9 p.m.
local time on December 31.

LOOKING NORTH

Around northern midwinter, the spectacular star clusters of Taurus lie due south (right), with Auriga and its bright star Capella high above them. To the southeast, the hunter Orion, accompanied by his dogs Canis Major and Canis Minor, rises to confront the mighty bull.

LOOKING SOUTH

JANUARY

These charts show the southern hemisphere skies during January—they are correct for 11 p.m. local time (midnight if Daylight Saving Time is in effect) on January 1, 10 p.m. local time on January 15, and 9 p.m. local time on January 31.

LOOKING NORTH

The two brightest stars in the sky, Sirius in Canis Major and Canopus in Carina, lie high overhead on summer nights, while the third brightest, Alpha Centauri, hangs in the southern sky (right). Orion and his retinue, meanwhile, are suspended over the northern horizon (left).

20°S **40°S** 60°S

LEPUS

CANIS MAJOR

COLUMBA

ERIDANUS

PUPPIS

CAELUM

HOROLOGIUM

FORNAX

PYXIS

DORADO

CETUS

PICTOR

RETIC-
ULUM

VOLANS

MENSA

ANTLIA

CHAMAELEON

HYDRUS

VELA

PHOENIX

HYDRA

CARINA

OCTANS

CRUX

MUSCA

TUCANA

APUS

GRUS

CENTAURUS

TRIANGULUM
AUSTRALE

PAVO

PISCIS
AUSTRINUS

SCULPTOR

CRATER

CIRCINUS

INDUS

CORVUS

NORMA

ARA

TELESCOPIUM

E

MICROSC-
OPIUM

CAPRICORNUS

W

LUPUS

CORONA
AUSTRALIS

SAGITTARIUS

SCORPIUS

LOOKING SOUTH

283

FEBRUARY

These charts show the southern hemisphere skies during February—they are correct for 11 p.m. local time (midnight if Daylight Saving Time is in effect) on February 1, 10 p.m. local time on February 14, and 9 p.m. local time on February 28.

CARINA

PICTOR

VELA

HOROLOGIUM

CAELUM

PUPPIS

ANTLIA

COLUMBA

PYXIS

CANIS MAJOR

FORNAX

LEPUS

HYDRA

CORVUS

ERIDANUS

MONOCEROS

CRATER

CANIS MINOR

SEXTANS

ORION

CANCER

GEMINI

LEO

TAURUS

AURIGA

LEO MINOR

CETUS

LYNX

VIRGO

ARIES

URSA MAJOR

COMA BERENICES

PERSEUS

CAMELO-PARDALIS

W

E

LOOKING NORTH

20°S **40°S** 60°S

The great celestial ship Argo Navis, composed of Puppis, Vela, and Carina, rides across southern skies on a broad sweep of the Milky Way, with the equally bright stars of Crux and Centaurus to its southeast (right). Looking north (left), the twin stars of Gemini are at their highest.

CANIS MAJOR

PYXIS

PUPPIS

LEPUS

COLUMBA

ANTLIA

HYDRA

VELA

CAELUM

PICTOR

CRATER

CARINA

HOROLOGIUM

ERIDANUS

VOLANS

DORADO

RETICULUM

CENTAURUS

CRUX

MUSCA

MENSA

FORNAX

CORVUS

CHAMAELEON

OCTANS

HYDRUS

CIRCINUS

APUS

TRIANGULUM

AUSTRALE

LUPUS

TUCANA

PHOENIX

NORMA

PAVO

INDUS

SCULPTOR

CETUS

VIRGO

ARA

TELESCOPIUM

GRUS

E

W

LIBRA

MICROSCOPIUM

PISCIS

AUSTRINUS

SCORPIUS

CORONA

AUSTRALIS

LOOKING SOUTH

MARCH

These charts show the southern hemisphere skies during March—they are correct for 11 p.m. local time (midnight if Daylight Saving Time is in effect) on March 1, 10 p.m. local time on March 15, and 9 p.m. local time on March 31.

LOOKING NORTH

Approaching the southern fall equinox, the prominent figure of Leo hangs upside down on the northern horizon, with Hydra and its smaller companion winding across the sky above it (left). The bright constellations Puppis, Vela, and Carina still dominate southern skies (right).

LOOKING SOUTH

APRIL

These charts show the southern hemisphere skies during April—they are correct for 11 p.m. local time (midnight if Daylight Saving Time is in effect) on April 1, 10 p.m. local time on April 15, and 9 p.m. local time on April 30.

LOOKING NORTH

Looking north (left) brilliant Spica and Arcturus
are the two most prominent stars in the sky,
while the rich galaxy clusters of Virgo and
Coma Berenices lie near the horizon.
To the south (right) the Milky Way
arcs from east to west across
the sky, with Centaurus and
Crux directly overhead.

CRATER

CORVUS

HYDRA

ANTLIA

CENTAURUS

VELA

PYXIS

CRUX

LIBRA

LUPUS

PUPPIS

MUSCA

NORMA

CIRCINUS

CARINA

TRIANGULUM
AUSTRALE

APUS

CHAMAELEON

VOLANS

COLUMBA

SCORPIUS

MENSA

CANIS
MAJOR

ARA

OCTANS

PICTOR

PAVO

DORADO

LEPUS

CORONA
AUSTRALIS

RETICULUM

CAELUM

OPHIUCHUS

TELESCOPIUM

HYDRUS

ORION

SAGITTARIUS

TUCANA

HOROLOGIUM

SERPENS
CAUDA

INDUS

GRUS

SCUTUM

FORNAX

ERIDANUS

MICRO-
SCOPIUM

PISCIS
AUSTRINUS

PHOENIX

SCULPTOR

E

W

LOOKING SOUTH

MAY

These charts show the southern hemisphere skies during May—they are correct for 11 p.m. local time (midnight if Daylight Saving Time is in effect) on May 1, 10 p.m. local time on May 15, and 9 p.m. local time on May 31.

CRUX

LUPUS

NORMA

CENTAURUS

VELA

SCORPIUS

ANTLIA

LIBRA

CRATER

SERPENS CAUDA

PYXIS

HYDRA

CORVUS

VIRGO

OPHIUCHUS

SCUTUM

SEXTANS

SERPENS CAPUT

COMA BERENICES

LEO

CORONA BOREALIS

BOOTES

AQUILA

LEO MINOR

CANES VENATICI

HERCULES

W

CANCER

E

URSA MAJOR

DRACO

LOOKING NORTH

20°S **40°S** 60°S

As Scorpius and Sagittarius rise through the eastern sky, they bring with them some of the richest regions of the Milky Way. Meanwhile, the far northern constellations of Hercules and Corona Borealis skim the horizon to the north (left), and Centaurus and Crux dominate the south (right).

LIBRA

CORVUS

CRATER

CENTAURUS

SCORPIUS

LUPUS

HYDRA

NORMA

CRUX

OPHIUCHUS

ANTLIA

TRIANGULUM
AUSTRALE

CIRCINUS

SERPENS
CAUDA

ARA

APUS

MUSCA

VELA

SCUTUM

CORONA
AUSTRALIS

TELESCOPIUM

PYXIS

PAVO

OCTANS

CHAMAELEON

CARINA

MENSA

VOLANS

PUPPIS

HYDRUS

SAGITTARIUS

INDUS

RETICULUM

TUCANA

PICTOR

CANIS
MINOR

MICROSCOPIUM

DORADO

AQUILA

GRUS

COLUMBA

CANIS
MAJOR

E

CAPRICORNUS

HOROLOGIUM

CAELUM

W

PHOENIX

LEPUS

PISCIS
AUSTRINUS

SCULPTOR

FORNAX

ERIDANUS

LOOKING SOUTH

JUNE

These charts show the southern hemisphere skies during June—they are correct for 11 p.m. local time (midnight if Daylight Saving Time is in effect) on June 1, 10 p.m. local time on June 15, and 9 p.m. local time on June 31.

CIRCINUS
ARA
LUPUS
NORMA
CENTAURUS
TELESCOPIUM
CORONA AUSTRALIS
SCORPIUS
HYDRA
SAGITTARIUS
LIBRA
CORVUS
CAPRICORNUS
SERPENS CAUDA
CRATER
OPHIUCHUS
SCUTUM
SERPENS CAPUT
AQUILA
VIRGO
CORONA BOREALIS
HERCULES
SEXTANS
COMA BERENICES
VULPECULA
CANES VENATICI
BOOTES
LYRA
SAGITTA
DELPHINUS
LEO
CYGNUS
EQUULEUS
LEO MINOR
URSA MAJOR
DRACO

W

E

LOOKING NORTH

20°S **40°S** 60°S

Around southern midwinter, brilliant red Antares in Scorpius lies overhead, while Triangulum Australe hovers over the south celestial pole (right). To the north, the bright stars Altair in Aquila and Vega in Lyra are rising (left), as Spica and Arcturus, in Virgo and Bootes, sink into the west.

LIBRA

OPHIUCHUS

SCORPIUS

VIRGO

LUPUS

NORMA

CENTAURUS

CORVUS

TELESCOPIUM

ARA

CORONA
AUSTRALIS

CIRCINUS

HYDRA

SAGITTARIUS

TRIANGULUM
AUSTRALE

CRUX

CRATER

PAVO

APUS

MUSCA

MICROSCOPIUM

OCTANS

CARINA

VELA

INDUS

CHAMAELEON

ANTLIA

MENSA

CAPRICORNUS

GRUS

TUCANA

VOLANS

HYDRUS

RETIC-
ULUM

EQUU-
LEUS

PHOENIX

PYXIS

PISCIS
AUSTRINUS

PICTOR

PUPPIS

AQUARIUS

SCULPTOR

HOROL-
OGIUM

DORADO

CAELUM

CANIS
MAJOR

FORNAX

COLUMBA

E

W

LOOKING SOUTH

JULY

These charts show the southern hemisphere skies
during July—they are correct for 11 p.m. local time
(midnight if Daylight Saving Time is in effect)
on July 1, 10 p.m. local time on July 15,
and 9 p.m. local time on July 31.

LOOKING NORTH

20°S **40°S** 60°S

From low southern latitudes, the northern "Summer Triangle" of Altair in Aquila, Vega in Lyra, and Deneb in Cygnus hangs inverted over the northern horizon, a broad arc of zodiac constellations sweeping above it (left). The richest parts of the Milky Way, in Sagittarius, lie directly overhead.

SERPENS
CAUDA

SCORPIUS

SAGITTARIUS

CORONA
AUSTRALIS

LIBRA

NORMA

LUPUS

TELESCOPIUM

CAPRICORNUS

MICROSCOPIUM

ARA

CIRCINUS

VIRGO

PISCIS
AUSTRINUS

PAVO

TRIANGULUM
AUSTRALE

INDUS

APUS

GRUS

CRUX

CENTAURUS

OCTANS

MUSCA

TUCANA

CHAMA-
ELEON

CORVUS

HYDRUS

CARINA

PHOENIX

MENSA

SCULPTOR

CRATER

RETICULUM VOLANS

VELA

AQUARIUS

HORO-
LOGIUM

PICTOR

ANTLIA

DORADO

HYDRA

E

FORNAX

CAELUM

PUPPIS

PYXIS

W

CETUS

COLUMBA

ERIDANUS

LOOKING SOUTH

AUGUST

These charts show the southern hemisphere skies
during August—they are correct for 11 p.m. local
time (midnight if Daylight Saving Time is
in effect) on August 1, 10 p.m. local time
on August 15, and 9 p.m. local
time on August 31.

LOOKING NORTH

Looking north, the Square of Pegasus is rising out of the east, with the small constellations of Sagitta, Equuleus, and Delphinus running before it (left). The bright constellation of Cygnus is at its highest above the northern horizon as Hercules sinks out of view.

CAPRICORNUS

SAGITTARIUS

MICROSCOPIUM

PISCIS
AUSTRINUS

AQUARIUS

CORONA
AUSTRALIS

GRUS

INDUS

TELESCOPIUM

SCORPIUS

SCULPTOR

ARA

NORMA

PAVO

TRIANGULUM
AUSTRALE

LIBRA

OCTANS

CIRCINUS

TUCANA

APUS

LUPUS

PHOENIX

HYDRUS

MUSCA

CHAMA-
ELEON

CRUX

MENSA

VOLANS

CARINA

CENTAURUS

HOROL-
OGIUM

RETICULUM

CETUS

FORNAX

DORADO

HYDRA

VELA

VIRGO

PICTOR

CAELUM

ANTLIA

CRATER

CORVUS

ERIDANUS

COLUMBA

PUPPIS

E

W

LOOKING SOUTH

SEPTEMBER

These charts show the southern hemisphere skies
during September—they are correct for 11 p.m.
local time (midnight if Daylight Saving Time
is in effect) on September 1, 10 p.m. local
time on September 15, and 9 p.m.
local time on September 30.

INDUS

TELESCOPIUM

ARA

GRUS

PHOENIX

CORONA
AUSTRALIS

MICROSCOPIUM

SCULPTOR

SAGITTARIUS

PISCIS AUSTRINUS

SCORPIUS

FORNAX

CAPRICORNUS

SERPENS
CAUDA

SCUTUM

AQUILA

AQUARIUS

EQUULEUS

CETUS

DELPHINUS

PEGASUS

SAGITTA

PISCES

OPHIUCHUS

VULPECULA

SERPENS
CAPUT

HERCULES

LYRA

CYGNUS

LACERTA

TRIANG-
ULUM

W

E

ANDROMEDA

ARIES

CEPHEUS

CASSIOPEIA

DRACO

LOOKING NORTH

As Cetus, with its renowned variable star Mira, rises from the east into northern skies (left), the rich Milky Way starfields of Ophiuchus, Sagittarius, and Scorpius descend into the west. Bright Fomalhaut in Piscis Austrinus hangs overhead, with the even more brilliant Achernar, in Eridanus, below it in southern skies (right).

AQUARIUS

CAPRICORNUS

PISCIS AUSTRINUS

MICROSCOPIUM

SCULPTOR

SAGITTARIUS

SCUTUM

GRUS

INDUS

PHOENIX

CORONA AUSTRALIS

SERPENS CAUDA

TUCANA

TELESCOPIUM

OPHIUCHUS

CETUS

PAVO

FORNAX

HYDRUS

TRIANGULUM AUSTRALE

ARA

SCORPIUS

HORO-LOGIUM

OCTANS

APUS

NORMA

RETICULUM

MENSA

CHAMA-ELEON

CIRCINUS

VOLANS

MUSCA

CAELUM

DORADO

CRUX

LUPUS

ERIDANUS

PICTOR

CARINA

LIBRA

COLUMBA

CENTAURUS

PUPPIS

E

W

LEPUS

VELA

HYDRA

CANIS MAJOR

PYXIS

ANTLIA

LOOKING SOUTH

OCTOBER

These charts show the southern hemisphere skies during October—they are correct for 11 p.m. local time (midnight if Daylight Saving Time is in effect) on October 1, 10 p.m. local time on October 15, and 9 p.m. local time on October 31.

20°S **40°S** 60°S

The area around the Square of Pegasus now hangs above the northern horizon, offering the chance to see the bright Andromeda and Triangulum galaxies (left). Looking south, skies are mostly barren with the bright stars of the southern Milky Way hugging the horizon (right).

LOOKING SOUTH

NOVEMBER

These charts show the southern hemisphere skies during November—they are correct for 11 p.m. local time (midnight if Daylight Saving Time is in effect) on November 1, 10 p.m. local time on November 15, and 9 p.m. local time on November 30.

LOOKING NORTH

As the brilliant stars of Taurus and Orion rise into northeastern skies, they bring a number of beautiful deep-sky objects with them (left). Aries and Perseus skim the horizon due north, while brilliant Achernar lies due south above the celestial pole (right).

20°S **40°S** 60°S

ERIDANUS

FORNAX

SCULPTOR

AQUARIUS

PHOENIX

PISCIS AUSTRINUS

HOROLOGIUM

CAELUM

GRUS

CAPRICORNUS

TUCANA

LEPUS

COLUMBA

PICTOR

RETICULUM

HYDRUS

DORADO

INDUS

CANIS MAJOR

OCTANS

MICROSCOPIUM

VOLANS

MENSA

CHAMAELEON

PAVO

SAGITTARIUS

CARINA

APUS

PUPPIS

TRIANGULUM AUSTRALE

MUSCA

TELESCOPIUM

CORONA AUSTRALIS

CIRCINUS

ARA

MONOCEROS

VELA

CRUX

NORMA

SCORPIUS

AQUILA

PYXIS

SCUTUM

ANTLIA

CENTAURUS

E

W

HYDRA

LUPUS

LOOKING SOUTH

DECEMBER

These charts show the southern hemisphere skies during December—they are correct for 11 p.m. local time (midnight if Daylight Saving Time is in effect) on December 1, 10 p.m. local time on December 15, and 9 p.m. local time on December 31.

20°S **40°S** 60°S

Taurus now lies due north, with Auriga and its bright star Capella on the horizon beneath it and the bright stars of Orion, Gemini, Canis Major, and Canis Minor to the east (left). Meanwhile, the equally brilliant constellations of Carina, Vela, and Puppis are returning to prominence in the southeast (right).

E

W

LOOKING SOUTH

GLOSSARY

Active galaxy
A galaxy that emits large amounts of energy from its central regions, probably generated as matter falls into a supermassive black hole at the heart of the galaxy.

Asteroid
One of the countless rocky worlds of the inner Solar System, largely confined in the main Asteroid Belt beyond the orbit of Mars.

Astronomical unit
A unit of measurement widely used in astronomy, equivalent to Earth's average distance from the Sun – roughly 93 million miles (150 million km).

Atmosphere
A shell of gases held around a planet or star by its gravity.

Atom
A basic building block of matter consisting of a central positively charged nucleus orbited by one or more negatively charged electrons, resulting in an overall neutral electric charge. The precise number of subatomic particles within an atom determines which element it forms.

Barred spiral galaxy
A spiral galaxy in which the arms are linked to the hub by a straight bar of stars and other material.

Bayer letters
Greek letters assigned to the brighter stars in a constellation, offering a general indication of their order of brightness.

Binary star
A pair of stars in orbit around one another. Because the stars in a binary pair were usually born at the same time, they allow a direct comparison of the way that stars with different properties evolve.

Black hole
A superdense point in space, usually formed by a collapsing stellar core of at least five solar masses. Its gravity is so powerful that even light cannot escape it.

Brown dwarf
A so-called "failed star" that lacks the mass to trigger fusion of hydrogen in its core, and so never starts to shine properly. Instead, brown dwarfs radiate low-energy radiation (mostly infrared) through gravitational contraction and a more limited form of fusion.

Celestial equator
An imaginary line around the sky lying directly above Earth's equator, and separating the sky into northern and southern hemispheres.

Celestial pole
A point in the sky that happens to lie directly above one of Earth's poles of rotation. As a result, the celestial poles remain stationary in the sky as Earth rotates, while all other points in the sky seem to circle around them.

Celestial sphere
An imaginary spherical shell around the Earth that provides a useful "model" for mapping the sky. The celestial equator and celestial poles are defined on the sphere in relation to Earth's own equator and poles.

Circumpolar
Descriptive of a constellation that lies close to either the north or south celestial pole, and as a result never sets in the skies across a large part of a particular hemisphere. Instead, circumpolar constellations simply rotate around the celestial pole in the course of each day.

Comet
A small chunk of rock and ice that typically orbits at the outer edges of the Solar System. When comets are disturbed, they often fall into elliptical orbits that bring them closer to the Sun. Here, the effects of solar radiation cause their surface ices to evaporate, enveloping the solid nucleus in a tenuous coma and a tail.

Constellation
Technically, one of 88 areas of the sky defined by boundary lines in right ascension and declination. Traditionally, however, constellations are patterns in the sky produced by joining certain stars with imaginary lines.

Dark nebula
A cloud of interstellar gas and dust that absorbs light, and becomes visible only when silhouetted against a field of stars or other nebulae.

Declination (Dec.)
A measurement in the equatorial coordinate system, roughly analogous to latitude on Earth's surface, used in conjunction with right ascension.

Dwarf planet

An object orbiting the Sun that has sufficient gravity to pull itself into a spherical shape, but which, unlike a true planet, has not cleared the region around it of other objects. Currently there are three known dwarf planets—the asteroid Ceres and the Kuiper Belt objects Pluto and Eris.

Eclipse

See solar eclipse, lunar eclipse.

Eclipsing binary

A binary star system in which one star regular passes in front of the other, causing the system's combined brightness to drop.

Ecliptic

A path around the sky describing the apparent motion of the Sun, produced as Earth orbits around it over the course of a complete year. The ecliptic marks the center line of the region where the planets and Moon are typically found. It passes through the 12 traditional zodiac constellations, and also through Ophiuchus.

Electromagnetic radiation

A form of energy that consists of combined electric and magnetic waves, and is able to propagate itself across a vacuum at the speed of light. The energy or temperature of an object emitting radiation affects the frequency and wavelength of the waves, which are reflected by other characteristics.

Elliptical galaxy

A galaxy consisting of stars in orbits that have no particular orientation, and generally lacking in star-forming gas. Ellipticals include both the smallest and the largest galaxies known, and are typically composed of long-lived, low-mass red and yellow stars.

Elliptical orbit

An orbit in the form of a stretched circle, with one axis (the "semi-major" axis) typically longer than the other. Two points called foci lie on the semi-major axis to either side of the ellipse's center, with the body being orbited located at one focus.

Emission nebula

A cloud of gas in space that glows at very specific wavelengths, producing a spectrum full of "emission lines." These nebulae are usually energized by the high-energy light of nearby stars, and are often associated with star-formation regions.

Equatorial coordinates

The most widely used astronomical coordinate system, measuring the locations of celestial objects in terms of their declination relative to the celestial equator, and their right ascension relative to the First Point of Aries.

First Point of Aries

The point on the celestial sphere marking the intersection of the ecliptic with the celestial equator at the northern vernal (spring) equinox. The First Point of Aries marks the point where the Sun moves from the southern hemisphere into the northern half of the sky, and is traditionally used as the origin of the equatorial coordinate system, marking 0h right ascension.

Flamsteed numbers

A system of numbers assigned to the fainter naked-eye stars within constellations (those not given Bayer letters). In general, Flamsteed numbers increase from east to west across a constellation.

Galaxy

An independent system of stars, gas, and other material with a size measured in thousands of light years, containing anything from millions to billions of stars.

Giant planet

A planet comprising a huge envelope of gas, liquid, or slushy ice (various frozen chemicals), perhaps around a relatively small rocky core.

Globular cluster

A dense ball of ancient and long-lived stars, typically found in orbit around a galaxy such as the Milky Way.

Horizontal coordinates

An *ad hoc* astronomical coordinate system in which the positions of objects are defined in terms of their altitude above the horizon and their "azimuth" or angle from due North. An object's horizontal coordinates are unique to a specific time and location on Earth, and change rapidly.

Infrared

Electromagnetic radiation with slightly less energy than visible light. Infrared radiation is typically emitted by warm objects too cool to glow visibly.

Irregular galaxy

A galaxy with no obvious structure, generally rich in gas, dust, and star-forming regions.

Kuiper Belt

A doughnut-shaped ring of icy worlds directly beyond the orbit of Neptune. The largest known Kuiper Belt Objects are Pluto and Eris.

Light year

A common unit of astronomical measurement, equivalent to the distance traveled by light in one year—roughly 5.9 trillion miles (9.5 million million km).

Luminosity

A measure of a star's energy output. Luminosity is technically measured in watts, but stars are so luminous that it is often simpler to compare them with the Sun. A star's luminosity in visible light is not necessarily equivalent to its entire energy output.

Lunar eclipse

An event during which the full Moon passes through Earth's shadow and is blocked from the illumination of direct sunlight.

Magnitude

Apparent magnitude is a measure of a star's brightness as seen from Earth, using a simple number—the lower the number, the brighter the object. The brightest star in the sky, Sirius, shines at magnitude –1.4, while the faintest naked-eye stars are around magnitude 6.0. A magnitude difference of 5.0 indicates that one star is perceived as 100 times brighter than another. Absolute magnitude is a system that estimates a star's magnitude as seen from a distance of 32.6 light years.

Main sequence

A term used to describe the longest phase in a star's life, during which it is relatively stable and shines by fusing hydrogen (the lightest element) into helium (the next lightest) at its core. During this period, the star obeys a general relationship that links its mass, size, luminosity, and color.

Multiple star

A system of two or more stars in orbit around a shared "center of gravity" (pairs of stars are also called binaries). Most of the stars in our galaxy are members of multiple systems rather than individuals like the Sun. However the vast majority of these systems are so distant or so tightly bound in orbit around each other that they are indivisible through the most powerful telescopes—instead, their multiple nature can only be discovered by studies of spectra that reveal their back-and-forth motion in orbit.

Nebula

Any cloud of gas or dust floating in space. Nebulae are the material from which stars are born, and into which they are scattered again at the end of their lives. They include emission nebulae, reflection nebulae, dark nebulae, and planetary nebulae. The word "nebula" means cloud in Latin, and was originally applied to any fuzzy object in the sky—including some now known to be star clusters or distant galaxies.

Neutron star

The collapsed core of a massive star, formed in a supernova explosion. A neutron star consists of compressed subatomic particles, and is the densest known object—although in the most massive stars, the core can collapse past this stage to form a black hole. Many neutron stars initially behave as pulsars.

Nova

A binary star system prone to dramatic outbursts, in which a white dwarf is pulling material from a companion star, building up a layer of gas around itself that then burns away in a violent nuclear explosion.

Nuclear fusion
The joining-together of light atomic nuclei (the central cores of atoms) to make heavier ones at very high temperatures and pressures, releasing excess energy in the process. Fusion is the process by which the stars shine, and the fusion of the lightest element, hydrogen, into the next lightest, helium, is its most efficient form.

Oort Cloud
A spherical shell of dormant comets, up to two light years across, surrounding the entire Solar System.

Open cluster
A large group of bright young stars that have recently been born from the same star-forming nebula, and may still be embedded in its gas clouds.

Orbit
The typically elliptical path that a smaller body follows around a more massive one under the influence of gravity. Circular orbits are just an unusual form of ellipse.

Planet
A world that follows its own orbit around the Sun, is massive enough to pull itself into a spherical shape, and which has cleared the space around it of other objects (apart from satellites). According to this definition, there are eight planets—Mercury, Venus, Earth, Mars, Jupiter, Saturn, Uranus, and Neptune.

Planetary conjunction
An apparent close encounter or alignment between planets in Earth's skies, created when planets just happen to lie in the same direction as seen from Earth, and possible because the orbits of the planets are all roughly aligned with the ecliptic (the plane of Earth's orbit around the Sun).

Planetary nebula
An expanding cloud of glowing gas sloughed off from the outer layers of a dying red giant star as it transforms into a white dwarf stellar remnant.

Pole star
A star that happens to lie close to one or other celestial pole, and which therefore remains more or less fixed in the sky. At present, Polaris in Ursa Minor forms the northern pole star, while the southern hemisphere has no bright equivalent. The effects of precession cause the stars to move slowly relative to the celestial pole, and so pole stars change over time.

Precession
A gradual rotation of the celestial sphere relative to the stars, created as tidal forces from the Sun and Moon cause Earth's axis of rotation to slowly change its orientation in space. Precession causes the First Point of Aries to move slowly westwards around the sky and the celestial poles to "wander" among the stars. As a result, objects slowly change their coordinates in right ascension and declination.

Protostar
A star that is still coalescing from the collapse of a gas cloud under its own gravity. As the center of the cloud heats up, it may emit infrared radiation.

Pulsar
A rapidly spinning neutron star (collapsed stellar core) with an intense magnetic field that channels its radiation out along two narrow beams that sweep across the sky like a cosmic lighthouse.

R.A.
See Right ascension

Radio waves
The lowest-energy form of electromagnetic radiation, with the longest wavelengths. Radio waves are emitted by cool gas clouds in space, but also by violent active galaxies and pulsars.

Red dwarf
A star with considerably less mass than the Sun—small, faint and with a low surface temperature. Red dwarfs fuse hydrogen into helium in their cores very slowly and live for much longer than the Sun, despite their size.

Red giant
A star passing through a phase of its life where its luminosity has increased hugely, causing its outer layers to expand and its surface to cool. Stars usually enter red giant phases when they exhaust fuel supplies in their core.

Reflection nebula
A cloud of interstellar gas and dust that shines as it reflects or scatters light from nearby stars. Reflection nebulae are typically bluish in color.

Right ascension (R.A.)
A measurement of celestial coordinates roughly analogous to longitude on Earth's surface, used in conjunction with declination. Right ascension is measured in hours, minutes, and seconds, and indicates an object's location relative to the First Point of Aries, which defines 0h right ascension.

Rocky planet
A relatively small planet composed largely of rocks and minerals, perhaps with a thin envelope of gas and liquid. Earth is the largest of the Solar System's four rocky planets.

Solar eclipse
A rare event during which the Moon passes directly across the face of the Sun, casting its shadow onto the surface of the Earth. The alignment must be so precise that a dramatic total eclipse is usually only visible in a narrow band across Earth's surface, but partial eclipses can be seen both more widely and more frequently.

Spectrum
The spread-out band of light created by passing light through a prism or similar device. The prism bends light by different amounts depending on its wavelength and color, so the spectrum reveals the precise intensities of light at different wavelengths.

Spiral galaxy
A galaxy consisting of a hub of old yellow stars, surrounded by a flattened disk of younger stars, gas, and dust, with spiral arms marking regions of current star formation.

Supernova
An enormous stellar explosion marking the death of a star far

Supernova remnant
A cloud of shredded, superhot gas that marks the aftermath of a supernova explosion. The term may also be used to describe the strange objects left behind in the wreckage by the collapsed stellar core—either a neutron star or a black hole.

Variable star
A star that changes its apparent brightness over time. Variations may be small or large, and may happen over very short or very long periods. They may be caused by interacting systems (as in eclipsing binaries and novae) or by genuine pulsations in the size and luminosity of an individual star.

Zodiac
A band of a dozen ancient constellations marking the Sun's annual path around the sky (the ecliptic). The planets and Moon are also usually found within the zodiac constellations, but thanks to the effects of precession, the ecliptic now also passes through a thirteenth constellation, the Serpent Bearer Ophiuchus.

MAJOR METEOR SHOWERS

Several showers of meteors or shooting stars make predictable returns every year. Each is named after the constellation from which it seems to radiate (see page 255), and most can be linked to a parent object that has left a trail of interplanetary debris along its orbit.

NAME	DATES	PEAK	RADIANT R.A.	RADIANT DEC.	ZHR*	PARENT OBJECT
Quadrantids	Dec 28–Jan 12	Jan 4	15h	+49°	120	Unknown
Lyrids	Apr 16–25	Apr 22	18h	+34°	18	Comet Thatcher
Pi Puppids	Apr 15–28	Apr 23	07h	–45°	variable	Comet 26P/Grigg-Skjellerup
Eta Aquariids	Apr 19–May 28	May 5	23h	–1°	65	Comet 1P/Halley
Arietids	May 22–Jul 2	Jun 7	03h	+24°	54	Comet 96P/Machholz
June Bootids	Jun 22–Jul 2	Jun 27	15h	+48°	variable	Comet 7P/Pons-Winnecke
Southern Delta Aquariids	Jul 2–Aug 23	Jul 29	23h	–16°	16	Comet 96P/Machholz
Beta Cassiopeids	Jul 3–Aug 19	Jul 29	24h	+59°	10	Unknown
Alpha Capricornids	Jul 3–Aug 5	Jul 29	20h	–10°	5	Comet 169P/NEAT
Perseids	Jul 17–Aug 24	Aug 12	03h	+58°	100	Comet 109P/Swift-Tuttle
Kappa Cygnids	Aug 3–25	Aug 17	19h	+59°	3	Minor planet 2008 ED69
Alpha Aurigids	Aug 28–Sep 5	Aug 31	06h	+39°	6	Comet C/1911 N1 (Kiess)
Draconids	Oct 6–10	Oct 8	17h	+54°	variable	Comet 21P/Giacobini-Zinner
Southern Taurids	Sep 10–Nov 20	Oct 10	02h	+09°	5	Comet 2P/Encke
Orionids	Oct 2–Nov 7	Oct 21	06h	+16°	25	Comet 1P/Halley
Northern Taurids	Oct 20–Dec 10	Nov 12	04h	+22°	5	Minor planet 2004 TG10
Leonids	Nov 6–30	Nov 17	10h	+22°	15	Comet 55P/Tempel-Tuttle
Alpha Monocerotids	Nov 15–25	Nov 21	08h	+01°	variable	Unknown
Phoenicids	Nov 28–Dec 9	Dec 6	01h	–53°	variable	Comet D/1819 W1 (Blanpain)
Puppid-Velids	Dec 1–15	multiple	08h	–45°	10	Unknown (complex series of showers)
Geminids	Dec 7–17	Dec 13	07h	+33°	120	Minor planet 3200 Phaethon
Ursids	Dec 17–26	Dec 23	14h	+76°	10	Comet 8P/Tuttle

* ZHR = Zenithal Hourly Rate, the estimated rate at which meteors would be seen if the meteor shower's radiant was located directly overhead at the zenith.

INDEX

ACKNOWLEDGMENTS

2: Babak Tafreshi, TWAN/Science Photo Library; 10: Gordon Garradd/Science Photo Library; 14: Jason Auch; 21tl:NASA, ESA, I. de Pater and M. Wong (University of California, Berkeley); tr: NASA, ESA, and the Hubble Heritage (STScI/AURA)-ESA/Hubble Collaboration; bl: ESO; br:NASA, ESA, and the Hubble Heritage Team (STScI/AURA); 28: Babak Tafreshi, TWAN/Science Photo Library; 30: David Parker/Science Photo Library; 32: Credit: NASA, ESA, HEIC, and The Hubble Heritage Team (STScI/AURA) Acknowledgment: R. Corradi (Isaac Newton Group of Telescopes, Spain) and Z. Tsvetanov (NASA); 34: T.A. Rector (University of Alaska Anchorage) and WIYN/NOAO/AURA/NSF; 36: Image Data - Subaru Telescope (NAOJ), Hubble Legacy Archive; Processing - Robert Gendler; 38: NASA/DOE/Fermi LAT Collaboration, CXC/SAO/JPL-Caltech/Steward/O. Krause et al., and NRAO/AUI; 40: T. A. Rector & B. A. Wolpa, NOAO, AURA, NSF; 42l: Research by Kloppenborg et al Nature 464, 870-872 (8 April 2010)". Graphic by John D. Monnier, University of Michigan.; r: Adam Block/Mount Lemmon SkyCenter/University of Arizona; 43: Adam Block/NOAO/AURA/NSF; 44: R. Barrena and D. López (IAC).; 46l: ESO; r: Robert Williams and the Hubble Deep Field Team (STScI) and NASA; 47: NASA, ESA and the Hubble Heritage Team STScI/AURA). Acknowledgment: A. Zezas and J. Huchra (Harvard-Smithsonian Center for Astrophysics); 48l: NASA/JPL-Caltech/STScI/CXC/UofA/ESA/AURA/JHU; r: NASA, ESA and R. de Grijs (Inst. of Astronomy, Cambridge, UK); 49: NASA, ESA and the Hubble Heritage Team STScI/AURA). Acknowledgment: J. Gallagher (University of Wisconsin), M. Mountain (STScI) and P. Puxley (NSF).; 50: X-ray: NASA/CXC/ Univ. of Maryland/A.S. Wilson et al.; Optical: Pal. Obs. DSS; IR: NASA/JPL-Caltech; VLA: NRAO/AUI/ NSF; 52l: Credit for the NICMOS Image: NASA, ESA,

M. Regan and B. Whitmore (STScI), and R. Chandar (University of Toledo), Credit for the ACS Image: NASA, ESA, S. Beckwith (STScI), and the Hubble Heritage Team (STScI/AURA); r: H. Ford (JHU/STScI), the Faint Object Spectrograph IDT, and NASA; 53: NASA, ESA, S. Beckwith (STScI), and The Hubble Heritage Team STScI/AURA); 54: ESO/L. Calçada; 56: Jack Burgess/Adam Block/NOAO/AURA/NSF; 58: ESA/Hubble and NASA; 60: The Hubble Heritage Team (AURA/STScI/NASA); 62: ESO; 64: NASA, ESA, the Hubble Heritage (STScI/AURA)-ESA/Hubble Collaboration, and the Digitized Sky Survey 2. Acknowledgment: J. Hester (Arizona State University) and Davide De Martin (ESA/Hubble); 66l: NASA/ Marshall Space Flight Center; c: Image courtesy of NRAO/AUI ; r: Richard Yandrick (Cosmicimage.com); 67: T. A. Rector/University of Alaska Anchorage and WIYN/NOAO/AURA/NSF; 68l: Zachary Grillo & the ESA/ESO/NASA Photoshop FITS Liberator; r: T.A. Rector/University of Alaska Anchorage and NOAO/ AURA/NSF; 69: NASA/JPL-Caltech/L. Rebull (SSC/ Caltech); 70: Adam Block/NOAO/AURA/NSF; 72l: NASA/JPL-Caltech/UCLA; r: NASA, ESA and T. Lauer (NOAO/AURA/NSF); 73: Adam Evans; 74: Caltech, Palomar Observatory, Digitized Sky Survey; Courtesy: Scott Kardel; 76l: N.A.Sharp/NOAO/AURA/NSF; r: Jean-Charles Cuillandre (CFHT) & Giovanni Anselmi (Coelum Astronomia), Hawaiian Starlight; 77: NASA, ESA, NRAO and L. Frattare (STScI). Science Credit: X-ray: NASA/CXC/IoA/A.Fabian et al.; Radio: NRAO/ VLA/G. Taylor; Optical: NASA, ESA, the Hubble Heritage (STScI/AURA)-ESA/Hubble Collaboration, and A. Fabian (Institute of Astronomy, University of Cambridge, UK); 78: Gemini Observatory, GMOS Team; 80: Richard and Leslie Maynard/Adam Block/ NOAO/AURA/NSF; 82l: P.Massey (Lowell), N.King (STScI), S.Holmes (Charleston), G.Jacoby (WIYN) AURA/NSF ; r: NASA/JPL-Caltech; 83: NASA, Hui Yang

University of Illinois ODNursery of New Stars; 84: Eckhard Slawik/Science Photo Library; 86: NASA, ESA and AURA/Caltech; 87l: NASA/JPL-Caltech/J. Stauffer (SSC/Caltech); r: NASA and The Hubble Heritage Team (STScI/AURA) Acknowledgment: George Herbig and Theodore Simon (Institute for Astronomy, University of Hawaii); 88: Credits for Optical Image: NASA/HST/ASU/J. Hester et al. Credits for X-ray Image: NASA/CXC/ASU/J. Hester et al.; r: NASA/CXC/ASU/J. Hester et al.; 89: NASA, ESA, J. Hester and A. Loll (Arizona State University); 90: NASA, Andrew Fruchter and the ERO Team [Sylvia Baggett (STScI), Richard Hook (ST-ECF), Zoltan Levay (STScI)]; 92: Celestial Image Co./Science Photo Library; 94: MASIL Imaging Team; 96l: ESO; c: ESO/ Oleg Maliy; r: ESO; 97: NASA, ESA and the Hubble Heritage (STScI/AURA)-ESA/Hubble Collaboration. Acknowledgement: Davide De Martin and Robert Gendler; 98: NOAO/AURA/NSF and N.A. Sharp (NOAO); 100: NASA/ESA/Hubble Heritage Team (STScI/AURA); r: ESO; r: G. Fritz Benedict, Andrew Howell, Inger Jorgensen, David Chappell (University of Texas), Jeffery Kenney (Yale University), and Beverly J. Smith (CASA, University of Colorado), and NASA; 101: G. Fazio (Harvard-Smithsonian Astrophysical Observatory) L. Jenkins (Goddard Space Flight Center) A. Hornschemeier (Goddard Space Flight Center) B. Mobasher (Space Telescope Science Institute) D. Alexander (University of Durham, UK) F. Bauer (Columbia University); 102: NOAO/AURA/NSF; 104l: NASA and The Hubble Heritage Team (STScI/ AURA); r: NASA/JPL-Caltech/R. Kennicutt (University of Arizona) and the SINGS Team; 105: NOAO/AURA/ NSF; 106l: NASA/JPL-Caltech and The Hubble Heritage Team (STScI/AURA); r: X-ray: NASA/ UMass/Q.D.Wang et al.; Optical: NASA/STScI/AURA/ Hubble Heritage; Infrared: NASA/JPL-Caltech/Univ. AZ/R Kennicutt/SINGS Team ; 107: NASA and The

Hubble Heritage Team (STScI/AURA); 108: Image Credit: Lynette Cook; 110: NASA/JPL-Caltech/L. Allen (Harvard-Smithsonian CfA) & Gould's Belt Legacy Team; 112l: NASA and The Hubble Heritage Team (STScI/AURA); c: ESA/Hubble & NASA; r: NASA/ESA, Friendlystar; 113: NASA, J. English (U. Manitoba), S. Hunsberger, S. Zonak, J. Charlton, S. Gallagher (PSU), and L. Frattare (STScI); 114l: ESO; r: NASA, ESA, STScI, J. Hester and P. Scowen (Arizona State University); 115: NASA, ESA, and The Hubble Heritage Team (STScI/AURA); 116: NASA & ESA; 118: NASA, ESA, the Hubble Heritage (STScI/AURA)-ESA/ Hubble Collaboration, and A. Evans (University of Virginia, Charlottesville/NRAO/Stony Brook University); -; 119: Adam Block/Science Photo Library; 120: NASA, The Hubble Heritage Team (STScI/AURA); 122: NASA/WikiSky; 124: NASA, ESA, and the Hubble SM4 ERO Team; 126: Bruce Balick (University of Washington), Jason Alexander (University of Washington), Arsen Hajian (U.S. Naval Observatory), Yervant Terzian (Cornell University), Mario Perinotto (University of Florence, Italy), Patrizio Patriarchi (Arcetri Observatory, Italy), NASA; 128l: NASA/JPL-Caltech/Univ. of Arizona; r: ESO/ VISTA/J. Emerson. Acknowledgment: Cambridge Astronomical Survey Unit; 129: NASA, ESA, C.R. O"Dell (Vanderbilt University), M. Meixner and P. McCullough (STScI); 130: NASA/JPL-Caltech/C. Martin (Caltech)/M. Seibert(OCIW); 132: Nigel Sharp/ NOAO/AURA; 134l: Andrea Dupree (Harvard-Smithsonian CfA), Ronald Gilliland (STScI), NASA and ESA; r: NASA and The Hubble Heritage Team (STScI/ AURA). Acknowledgment: C. R. O"Dell (Vanderbilt University); 135l: ESO/J. Emerson/VISTA. Acknowledgment: Cambridge Astronomical Survey Unit; r: © Stocktrek Images/Corbis; 136l: NASA, ESA and L. Ricci (ESO); r: NASA; K.L. Luhman (Harvard-Smithsonian Center for Astrophysics,

Cambridge, Mass.); and G. Schneider, E. Young, G. Rieke, A. Cotera, H. Chen, M. Rieke, R. Thompson (Steward Observatory, University of Arizona, Tucson, Ariz.); 137: NASA,ESA, M. Robberto (Space Telescope Science Institute/ESA) and the Hubble Space Telescope Orion Treasury Project Team; 138: NASA, ESA and H.E. Bond (STScI); 140l: NASA/ESA and The Hubble Heritage Team (AURA/STScI).; r: Nick Wright (Nwright6302); 141: ESO; 142: NASA, H.E. Bond and E. Nelan (Space Telescope Science Institute, Baltimore, Md.); M. Barstow and M. Burleigh (University of Leicester, U.K.); and J.B. Holberg (University of Arizona); 144l: NASA/JPL-Caltech/Harvard-Smithsonian CfA; c: NASA, ESA, and R. Humphreys (University of Minnesota); r: ESO/B. Bailleul; 145: NASA and The Hubble Heritage Team (STScI); 146: NASA, ESA, the Hubble Heritage (STScI/AURA)-ESA/Hubble Collaboration, and W. Keel (University of Alabama); 148l: NASA and The Hubble Heritage Team (STScI/AURA). Acknowledgment: C. Conselice (U. Wisconsin/STScI); c: NASA/WikiSky; r: ESO; 149: NASA, ESA, and the Hubble Heritage Team (STScI/AURA). Acknowledgment: R. O¨Connell (University of Virginia) and the Wide Field Camera 3 Science Oversight Committee; 150: NASA, ESA, and the Hubble Heritage Team (STScI/AURA)-ESA/Hubble Collaboration; 152: Eckhard Slawik/Science Photo Library; 154l: ESO/WFI (Optical); MPIfR/ESO/APEX/A. Weiss et al. (Submillimetre); NASA/CXC/CfA/R.Kraft et al. (X-ray); r: NASA, ESA, and the Hubble Heritage (STScI/AURA)-ESA/Hubble Collaboration. Acknowledgment: R. O¨Connell (University of Virginia) and the WFC3 Scientific Oversight Committee; 155: ALMA (ESO/NAOJ/NRAO); ESO/Y. Beletsky; 156l: ESO; r: NASA and The Hubble Heritage Team (STScI/AURA). Acknowledgment: A. Cool (SFSU); 157: NASA, ESA, and the Hubble SM4 ERO Team; 158: NASA and The Hubble Heritage Team

(STScI/AURA). Acknowledgment: C.R. O¨Dell (Vanderbilt University); 160: N.A.Sharp, Mark Hanna, REU program/NOAO/AURA/NSF; 162l: NASA, The Hubble Heritage Team, STScI, AURA; c: NASA/Wikisky; r: Robert Gendler/Science Photo Library; 163: Royal Observatory, Edinburgh/AAO/Science Photo Library; 164: ESO; 166l: ESA/Hubble & NASA; r: ESO/INAF-VST/OmegaCAM. Acknowledgement: OmegaCen/Astro-WISE/Kapteyn Institute; 167: ESO; 168l: ESO/S. Guisard; r: NASA/CXC/MIT/F. Baganoff, R. Shcherbakov et al. ; 169: NASA/JPL-Caltech/ESA/CXC/STScI; 170: Hubble/Wikisky; 172: NASA, ESA and P. Kalas (University of California, Berkeley, USA); 174: ESO and Digitized Sky Survey 2. Acknowledgment: Davide De Martin; 176l: ESO/J. Emerson/VISTA. Acknowledgment: Cambridge Astronomical Survey Unit; c: ESO; r: ESO/R. Gendler; 177: NASA, ESA, S. Beckwith (STScI) and the HUDF Team; 178: NASA/JPL-Caltech/T. Pyle (SSC); 180: WikiSky; 182: NASA/JPL-Caltech/SSC; 184: NASA, ESA and Orsola De Marco (Macquarie University); 186: ESO; 188: ESO; 190l: NASA, ESA, and P. Hartigan (Rice University); r: The Hubble Heritage Team (STScI/AURA/NASA); 191: ESO; 192: NASA, ESA, and M. Livio and the Hubble 20th Anniversary Team (STScI); 194: NOAO/AURA/NSF; 195l: NASA, ESA, R. O¨Connell (University of Virginia), F. Paresce (National Institute for Astrophysics, Bologna, Italy), E. Young (Universities Space Research Association/Ames Research Center), the WFC3 Science Oversight Committee, and the Hubble Heritage Team (STScI/AURA); r: X-ray: NASA/CXC/CfA/M.Markevitch et al.; Optical: NASA/STScI; Magellan/U.Arizona/D.Clowe et al.; Lensing Map: NASA/STScI; ESO WFI; Magellan/U. Arizona/D.Clowe et al.; 196l: ESA/PACS/SPIRE/Thomas Preibisch, Universitäts-Sternwarte München, Ludwig-Maximilians-Universität München, Germany.; r: Credit for Hubble Image: NASA, ESA, N. Smith

(University of California, Berkeley), and The Hubble Heritage Team (STScI/AURA). Credit for CTIO Image: N. Smith (University of California, Berkeley) and NOAO/AURA/NSF; 197: ESO/T. Preibisch; 198: ESO/Y. Beletsky; 200: Raghvendra Sahai and John Trauger (JPL), the WFPC2 science team, andNASA/ESA; 202: NASA, Andrew S. Wilson (University of Maryland); Patrick L. Shopbell (Caltech); Chris Simpson (Subaru Telescope); Thaisa Storchi-Bergmann and F. K. B. Barbosa (UFRGS, Brazil); and Martin J. Ward (University of Leicester, U.K.); 204: X-ray: NASA/CXC/UVa/M. Sun et al; H-alpha/Optical: SOAR/MSU/NOAO/UNC/CNPq-Brazil/M.Sun et al.; 206: ESO; 208: NASA/WikiSky; 210: ESO; 212: NASA/JPL-Caltech/University of Virginia/R. Schiavon (Univ. of Virginia); 214l: ESA/Hubble (Davide De Martin), the ESA/ESO/NASA Photoshop FITS Liberator & Digitized Sky Survey 2; c: ESO; r: NASA and Ron Gilliland (Space Telescope Science Institute); 215: NASA, ESA, and G. Meylan (Ecole Polytechnique Federale de Lausanne); 216l: ESA/Hubble and Digitized Sky Survey 2. Acknowledgements: Davide De Martin (ESA/Hubble); r: NASA, ESA and A. Nota (ESA/STScI, STScI/AURA); 217: NASA, ESA, and the Hubble Heritage Team (STScI/AURA) - ESA/Hubble Collaboration; 218: NASA/WikiSky; 220: ESO/A.-M. Lagrange et al.; 222l: ESO; c: NASA, ESA, and the Hubble Heritage Team (STScI/AURA)-ESA/Hubble Collaboration. Acknowledgment: D. Gouliermis (Max Planck Institute for Astronomy, Heidelberg); r: ESO/R. Fosbury (ST-ECF); 223: NASA, ESA, F. Paresce (INAF-IASF, Bologna, Italy), R. O¨Connell (University of Virginia, Charlottesville), and the Wide Field Camera 3 Science Oversight Committee; 224: ESO; 226: ESO; 228: Alex Cherney, Terrastro.com/Science Photo Library; 230: F. Char/ESO; 234: Roger Ressmeyer/CORBIS; 236: George Post/Science Photo Library; 239bl: MESSENGER, NASA, JHU APL, CIW;

br: NASA/SOHO; 241bl: Brocken Inaglory; br: NASA/JPL; 243bl: NASA and The Hubble Heritage Team (STScI/AURA). Acknowledgment: J. Bell (Cornell U.), P. James (U. Toledo), M. Wolff (SSI), A. Lubenow (STScI), J. Neubert (MIT/Cornell); br: Larry W. Koehn/Science Photo Library; 244: NASA/JPL-Caltech/UCAL/MPS/DLR/IDA; 245bl: NASA, ESA, J. Parker (Southwest Research Institute), P. Thomas (Cornell University), L. McFadden (University of Maryland, College Park), and M. Mutchler and Z. Levay (STScI); br: NEAR Project, NLR, JHUAPL, Goddard SVS, NASA; 247bl: NASA/JPL; br: NASA/JPL; 249bl: NASA and The Hubble Heritage Team (STScI/AURA). Acknowledgment: R.G. French (Wellesley College), J. Cuzzi (NASA/Ames), L. Dones (SwRI), and J. Lissauer (NASA/Ames); br: NASA/JPL-Caltech/Space Science Institute; 251bl: Erich Karkoschka (University of Arizona) and NASA; br: NASA/JPL; 253bl: NASA/JPL; br: NASA/JPL/USGS; 254: ESO/Sebastian Deiries; 255bl: Gil-Estel; br: Dr Fred Espenak/Science Photo Library; 256: Ben Canales/Science Photo Library;

All other maps and illustrations by Pikaia Imaging www.pikaia-imaging.co.uk

Quercus

New York • London

© 2013 Quercus Editions Ltd

ISBN 978-1-62365-088-9

Library of Congress Control Number: 2013937740

Distributed in the United States and Canada by
Random House Publisher Services
c/o Random House, 1745 Broadway
New York, NY 10019

Manufactured in China

2 4 6 8 10 9 7 5 3 1

www.quercus.com